Modern Organic Chemistry

3rd Edition

A. ATKINSON

B.SC., A.R.I.C., B.SC. (ECON.), DIP.ED.

Formerly Senior Chemistry Master Wolverhampton and Stockton Grammar Schools, Principal High School Malacca and Victoria Institution Kuala Lumpur, Principal Kampala Technical College

Stanley Thornes (Publishers) Ltd

First published in 1973 by
Stanley Thornes (Publishers) Ltd, Old Station Drive, Leckhampton,
CHELTENHAM GL53 0DN.

Second edition 1977
Reprinted 1980 with corrections and new material

Third edition 1986

British Library Cataloguing in Publication Data
Atkinson, Arthur
Modern organic chemistry.—3rd ed.
1. Chemistry, Organic
I. Title
547 QD253

ISBN 0-85950-656-8

Acknowledgements

The Author and the Publishers wish to record their thanks to the University of London and the University of Cambridge for their permission to use examination questions reproduced in this book.

Typeset by Tech-Set, Gateshead, Tyne & Wear.
Printed and bound in Great Britain at The Bath Press, Avon.

Contents

Preface

This book is intended for students studying Organic Chemistry for G.C.E. ('A' level), and overseas H.S.C., and similar examinations. No previous knowledge of the subject is assumed. Emphasis is placed on functional groups of series of compounds and on their reactions; in this way facts which are apparently unrelated fit into a logical system. Various aspects of electronic valency theory and molecular structure are introduced early on, and others are discussed when appropriate.

Laboratory preparations of important compounds are given in detail. Practical details of simple reactions are given near the beginning of each chapter. Teachers can select those tests most suitable for their syllabus; the tests are too numerous to be done by all students in a two year course.

Names recommended by the 1985 report of the Association for Science Education have been used, but common current names are also quoted where this is thought desirable in the interests of clarity. Examination Boards will accept any meaningful name from students.

SI units, signs, symbols and abbreviations are used throughout, and additional 'labelling' information has been incorporated into the diagrams so as to enhance their descriptive value.

It is hoped that the approach to the subject is imaginative enough to encourage students to think that organic chemistry is in itself an intellectual challenge, and that through its understanding an insight into the very essence of life itself can be gained.

A.A.

Note on the Third Edition

The book has been improved by removing a few minor errors, adding several new diagrams showing modern apparatus, and by bringing the text into line with most modern 'A' level Chemistry syllabuses, including those of the Certificate of Sixth Year Studies, Scotland, and the Northern Ireland Schools Examination Council.

A completely new feature is the addition of a revision summary at the end of each chapter. A summary contains in the briefest form the essential factual chemistry in the chapter. It should reduce the time wasted in making notes and be most useful during revision before examinations. The tables in the summaries will be widely appreciated by both pupils and teachers.

1986

Chapter 1
STRUCTURES OF ORGANIC COMPOUNDS

Quantitative analysis of organic compounds

The elements present in an organic compound are determined by qualitative analysis. The percentages by mass of the elements are then determined and the empirical formula is calculated. The following outlines methods which have been used in quantitative analysis. The mass of organic compound is assumed to be a g in each experiment.

Carbon and hydrogen. Place the compound in a combustion tube. Heat in a slow stream of oxygen, and pass the products over hot copper (II) oxide to ensure complete oxidation. The products are water, from hydrogen, and carbon dioxide, from carbon. Absorb the water in calcium chloride tubes, and let the gain in mass be b g. Absorb the carbon dioxide in bulbs containing potassium hydroxide solution, and let the gain in mass be c g.

$$\begin{array}{cc} C \rightarrow CO_2 \\ 12\,g \quad 44\,g \end{array} \qquad\qquad \begin{array}{cc} 2H \rightarrow H_2O \\ 2\,g \quad 18\,g \end{array}$$

Since the mass of the organic compound was a g, the percentages by mass of hydrogen and carbon are:

$$\text{Hydrogen} = \frac{2b}{18} \times \frac{100}{a} = \frac{b}{9} \times \frac{100}{a}$$

$$\text{Carbon} = \frac{12c}{44} \times \frac{100}{a} = \frac{3c}{11} \times \frac{100}{a}$$

Nitrogen (by conversion to gaseous nitrogen). Mix the organic compound with excess copper(II) oxide and place in a combustion tube. Pass carbon dioxide to remove all air from the apparatus. Heat the mixture. Oxides of nitrogen are formed. Pass them over hot copper gauze, which reduces them to nitrogen, and collect this gas over concentrated potassium hydroxide. Pass carbon dioxide again to ensure no nitrogen remains in the combustion tube. Measure the volume of nitrogen, and convert it to s.t.p. Let the volume be b cm^3 at s.t.p.

$$\underset{28\,\text{g}}{2\text{N(in organic compound)}} \rightarrow \text{NO and NO}_2 \rightarrow \underset{22\,400\,\text{cm}^3}{\text{N}_2}$$

Mass of 22 400 cm^3 of nitrogen at s.t.p. is 28 g.

$\therefore$ Mass of b cm^3 of nitrogen at s.t.p. is $\frac{28b}{22\,400}$ g, and the percentage by mass of nitrogen is

$$\frac{28b}{22\,400} \times \frac{100}{a}$$

Nitrogen (by conversion to ammonia). Heat the compound with concentrated sulphuric acid to which potassium hydrogensulphate has been added to raise its boiling point. The nitrogen is converted to ammonium sulphate. Allow the mixture to cool, dilute, add excess sodium hydroxide, and distil the ammonia into a known volume (excess) of standard hydrochloric acid. Determine the amount of acid unused by titration against known alkali (as described in books of volumetric analysis) and calculate the acid required (b cm^3) to react with the ammonia.

$$\underset{14\,\text{g}}{\text{N (in organic compound)}} \rightarrow \text{NH}_3 \equiv \underset{1000\,\text{cm}^3\,\text{M}}{\text{HCl}}$$

1000 cm^3 M HCl $\equiv$ 14 g of nitrogen

$\therefore$ b cm^3 M HCl $\equiv \frac{14b}{1000}$ g of nitrogen, and the percentage by mass of nitrogen is

$$\frac{14b}{1000} \times \frac{100}{a}$$

This method is not suitable for nitro-compounds because all their nitrogen is not converted to ammonia.

Halogens. Heat the compound in a thick-walled sealed glass tube with excess fuming nitric acid and solid silver nitrate. Silver halide is precipitated. Cool the tube and break it under water. Filter the halide, wash, dry and weigh. Let the mass of silver halide be b g.

$$\underset{35.5\,g}{Cl} \rightarrow \underset{143.5\,g}{AgCl(s)} \qquad \underset{80\,g}{Br} \rightarrow \underset{188\,g}{AgBr(s)} \qquad \underset{127\,g}{I} \rightarrow \underset{235\,g}{AgI(s)}$$

143.5 g of silver chloride contain 35.5 g of chlorine.

$\therefore$ b g of silver chloride contain $\dfrac{35.5b}{143.5}$ g of chlorine,

and the percentage of chlorine is

$$\frac{35.5b}{143.5} \times \frac{100}{a}$$

The percentages of bromine and iodine are calculated similarly. (This method assumes that the organic compound contains one halogen only.)

Sulphur. The method is similar to that for halogens but silver nitrate is not used. Heat the organic compound with fuming nitric acid. Sulphur is converted to sulphuric acid. Allow the product to cool, dilute, and add barium nitrate solution. Barium sulphate is precipitated. Filter, wash, dry and weigh. Let the mass of barium sulphate be b g.

$$\underset{32\,g}{S} \text{ (in organic compound)} \rightarrow H_2SO_4 \rightarrow \underset{233\,g}{BaSO_4}$$

The percentage of sulphur is $\dfrac{32b}{233} \times \dfrac{100}{a}$.

Oxygen. Add all the percentages of the elements present. If the total is less than 100, the difference is assumed to be due to oxygen.

By modern methods and techniques it is possible to do a full quantitative analysis of an organic compound even if only 0.01 g is available.

Calculation of empirical formula

An organic compound contains 48.7 per cent of carbon and 8.1 per cent of hydrogen. What is its empirical formula?

Total percentage of carbon and hydrogen $= 48.7 + 8.1 = 56.8$
Percentage of oxygen $= 100 - 56.8 = 43.2$ per cent

	Carbon	Hydrogen	Oxygen
Percentage composition by mass	48.7	8.1	43.2
Relative number of atoms			
i.e. $\frac{\text{percentage composition}}{\text{relative atomic mass}}$	$\frac{48.7}{12}$	$\frac{8.1}{1}$	$\frac{43.2}{16}$
=	4.06	8.1	2.7
Divide by smallest number	$\frac{4.06}{2.7}$	$\frac{8.1}{2.7}$	$\frac{2.7}{2.7}$
=	1.5	3	1
Convert to whole numbers	3	6	2

The empirical formula is $C_3H_6O_2$

Determination of molecular formula

The relative molecular mass of an organic compound can be determined from (a) the density of its vapour, and (b) the elevation of the boiling point or depression of the freezing point of its solutions. The theory and methods are described in books on physical chemistry.

The formula $C_2H_3O_2$ indicates that the lowest possible relative molecular mass is $(2 \times 12) + 3 + (2 \times 16) = 59$. If the true relative molecular mass is 118, the molecular formula is double the empirical formula, that is, $(C_2H_3O_2)_2$ or $C_4H_6O_4$.

The percentage compositions of ethyne and benzene are the same: carbon, 92.3 per cent and hydrogen 7.7 per cent. Their empirical formulae are therefore the same: CH. The vapour density of ethyne is 13 and therefore its relative molecular mass is 26. Its molecular formula is $(CH)_2$ or C_2H_2. The relative molecular mass of benzene is 78 and its molecular formula is C_6H_6.

All the methods so far described in this chapter have been used for over a century. Four modern methods of obtaining information about the composition and structure of compounds are infrared absorption spectra, X-ray diffraction, electron diffraction and mass spectra. A brief account of each method follows.

Infrared absorption spectra

The wavelengths of infrared radiation vary from about 2500 nm to 25 000 nm, and they are longer than those of visible radiation or light. When passed through the vapour or liquid form of an organic compound, certain wavelengths are absorbed. The radiation absorbed is that which causes vibration of the atoms relative to each other. Each bond, e.g. C—H, N—H, C=O, and so on, absorbs radiation which is characteristic.

Reference books of recorded spectra are available. Their use enables the groups present in a molecule of an organic compound to be identified from its infrared absorption spectrum. A modern spectrometer produces and records a spectrum in a few minutes.

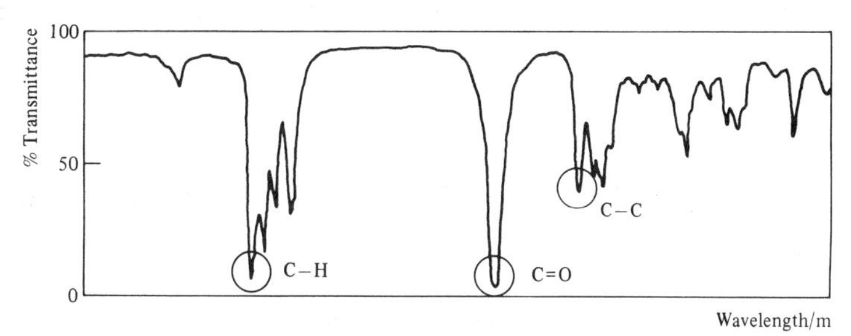

Fig. 1.1. Infrared absorption spectrum (general form)

Fig. 1.1 shows the general form of infrared spectra, in which the percentage of radiation transmitted is plotted against wavelength. The peaks extend from the top of the trace downwards, and pronounced dips indicate wavelengths at which strong absorption occurs. The wavelengths absorbed depend on the atoms joined to a bond and the nature of the bond itself (whether single, double or triple). If possible examine spectra of (a) hexane and hexene to show the effects of single and double bonds, (b) an alcohol and its ether isomer (C—O—H and C—O—C), (c) an aldehyde and a ketone, and (d) a carboxylic acid to show the effect of strong hydrogen bonds.

X-ray diffraction

The wavelengths of X-rays vary from about 10 nm to 0.001 nm, and are about 10 000 times smaller than those of light. X-ray wavelengths are much the same as the distances between atoms in crystals. X-rays are diffracted by particles (atoms or ions) of a solid crystal if they are arranged in a regular pattern. Diffracted X-rays form a regular pattern of spots when they fall on a photographic plate.

A single crystal can be replaced by a powder, and concentric circles are then formed on the photographic plate.

Distances between layers of molecules, atoms or ions in a crystal or powder can be determined from the photographs. Since the diffraction of X-rays is produced by the electrons around an atom or ion, a hydrogen atom produces practically no diffraction because it has only one electron. It is possible to calculate bond lengths and bond angles from the photographs; computers are used to deduce the detailed structure.

Electron diffraction

Electrons from an electrically heated filament are accelerated by an electric field. They then pass through a gas or vapour, which diffracts them. The diffracted electrons fall on a photographic plate and form rings of dark patches. Even hydrogen atoms can be located fairly well by this method. Bond lengths, bond angles, and structures of molecules can be deduced from the photographs.

Mass spectrometer (for relative molecular mass and structure)

A beam of high energy electrons is passed through vapour of the organic substance which is under very low pressure. Positive ions of the vapour are formed:

$$\text{Compound} + \text{fast electron} \rightarrow \text{Compound ion} + \text{two slow electrons}$$

The compound may also split into two or more parts and these also produce positive ions. The various ions have different charge/mass ratios and also have different speeds. For example, propane forms the ions: $C_3H_8^+$, $C_3H_8^{2+}$, $C_2H_5^+$ and CH_3^+. The last two are produced when C_3H_8 molecules split and then ionize.

$$C_3H_8 + e^-(\text{fast}) \rightarrow C_3H_8^+ + 2e^-(\text{slow})$$

The positive ions are accelerated by an electric field and they all move at about the same speed. The ions are then deflected by a magnetic field, and ions with the same charge/mass ratio come to one focus on a conductor. All other ions hit the walls of the spectrometer. Different sets of ions are brought to a focus by changing the magnetic field. The charge received by the conductor is measured by an electrometer connected to it.

The mass of each ion can be calculated from the known electric and magnetic fields. The relative abundances of the various ions are deduced from the currents produced in the detector.

The highest mass calculated is that of one ion of the compound and gives its relative molecular mass; lower masses are those of ions of fission products of the compound. The relative molecular mass helps to determine or confirm the molecular formula, and the various fission products may indicate the molecular structure. For example, a compound containing only carbon, hydrogen and oxygen and which has a relative molecular mass of 102 might have one of the following formulae: $C_4H_6O_3$, $C_5H_{10}O_2$, $C_6H_{14}O$. If the relative molecular mass was 200, a possible formula is $C_{10}H_{16}O_4$; try to write some of the other eight possible formulae using only carbon, hydrogen and oxygen.

Electronic structures of atoms

Atoms consist essentially of three fundamental particles called electrons, protons and neutrons. Protons and neutrons together are called nucleons, and they are in a tiny central *nucleus* which is surrounded by electrons. Properties of the particles are:

	Masses		*Charges*	
	Actual/kg	*Relative*	*Actual/C*	*Relative*
Electron	9.109×10^{-31}	0.000 55	-1.6×10^{-19}	−1
Proton	1.673×10^{-27}	1.007 4	$+1.6 \times 10^{-19}$	+1
Neutron	1.675×10^{-27}	1.008 9	0	0

The relative masses are on the carbon-12 scale.

A proton and an electron have equal and opposite charges. The masses of a proton and a neutron are about equal, while the mass of an electron is almost negligible. Therefore almost all the mass of an atom is concentrated in its nucleus, which is positively charged.

The number of protons in a nucleus is the *proton number* Z or *atomic number* of the element (and it is equal to the number of electrons). The number of neutrons in a nucleus is the *neutron number* N, and the number of nucleons is the *nucleon number* A. Therefore $A = Z + N$. The compositions of some atoms are:

	Atomic number	*Protons*	*Neutrons*	*Electrons*
Hydrogen	1	1	0	1
Carbon	6	6	6	6
Nitrogen	7	7	7	7
Oxygen	8	8	8	8
Fluorine	9	9	10	9
Sodium	11	11	12	11
Chlorine	17	17	18 or 20	17
Uranium	92	92	146	92

Electrons are distributed around a nucleus in definite energy levels or *shells*. In ordinary atoms, there are seven possible shells denoted by the numbers 1 to 7 or the letters K to Q. An electron in a K or 1 shell has least energy and the space it occupies, called its *orbital*, is nearest to the nucleus. The second or L shell is of higher energy and the orbital of an electron in it is further from the nucleus.

A hydrogen atom has one electron in the K shell and a helium atom has two, called a *duplet*. The electronic configurations of hydrogen and helium atoms are K1 and K2 respectively, usually written 1 and 2 without the K. The configurations of neon and argon are K2, L8 and K2, L8, M8 respectively, usually written 2,8 and 2,8,8. The eight electrons are called an *octet*. A duplet and octet take no part in ordinary chemical reactions. Other configurations are:

			Halogens	*Noble gases*	*Alkali metals*
				He 2	Li 2,1
C 2,4	N 2,5	O 2,6	F 2,7	Ne 2,8	Na 2,8,1
			Cl 2,8,7	Ar 2,8,8	K 2,8,8,1

Tetrahedral distribution of bonds

A covalent bond between two atoms consists of two shared electrons, one from each atom. A carbon atom has four electrons in its outer L shell and a hydrogen atom has one electron in its K shell. A carbon atom can combine with four hydrogen atoms. The carbon atom shares the electrons of the hydrogen atoms, and each hydrogen atom shares one electron of the carbon. Four covalent bonds join the carbon atom to the hydrogen atoms. The two orbitals of each shared electron pair overlap and form a single molecular orbital around the carbon and hydrogen nuclei and bond them together. In this way the carbon atom attains a stable neon octet in its L shell and each hydrogen atom attains a stable helium duplet:

```
      H                                   H
      |                                   ×·
  H — C — H            or            H ·× C ·× H
      |                                   ×·
      H                                   H
```

Each line (–) in the left formula represents two shared electrons. The dots and crosses represent electrons from different atoms but the electrons are identical and indistinguishable in the bonds.

The four covalent bonds of carbon are represented on paper at right angles to each other. If this arrangement is correct there should be two dichloromethanes CH_2Cl_2:

```
      Cl                                  H
      |                                   |
  H — C — H                           H — C — Cl
      |                                   |
      Cl                                  Cl
```

Only one compound exists.

Electron pair repulsion theory. The four electron pairs of a carbon atom in methane (and other alkane molecules) repel each other and are as far apart as possible. This is so when the carbon nucleus is at the centre of a regular tetrahedron and the four bonds are directed towards the four corners. The angle between two adjacent bonds is $109° 28'$. The four bonds are distributed symmetrically in space.

Fig. 1.2 shows the relation of the four hydrogen atoms to the carbon atom and the tetrahedral bonds in methane. All structural formulae of alkanes must be imagined with each carbon atom surrounded by four other atoms arranged symmetrically.

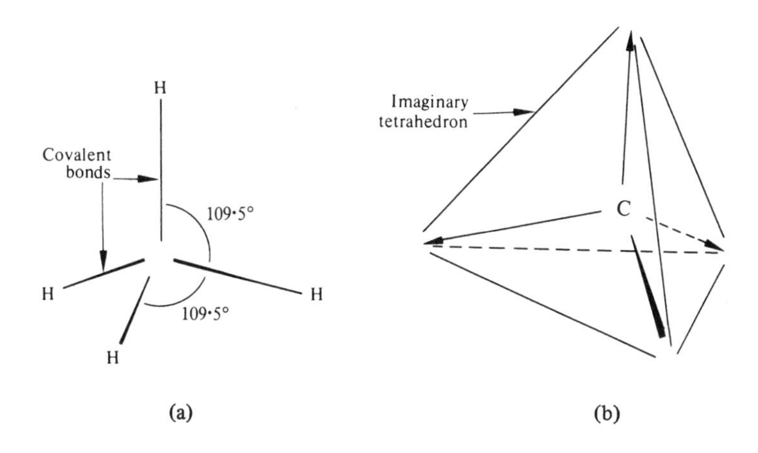

Fig. 1.2. Tetrahedral structure of methane

Fig. 1.3 shows two chlorine atoms and two hydrogen atoms bonded to the same carbon atom, forming a molecule of dichloromethane, CH_2Cl_2. The two configurations of the four atoms may appear different at first glance, but they are identical. Make ball-and-spring models of a molecule to confirm that this is so.

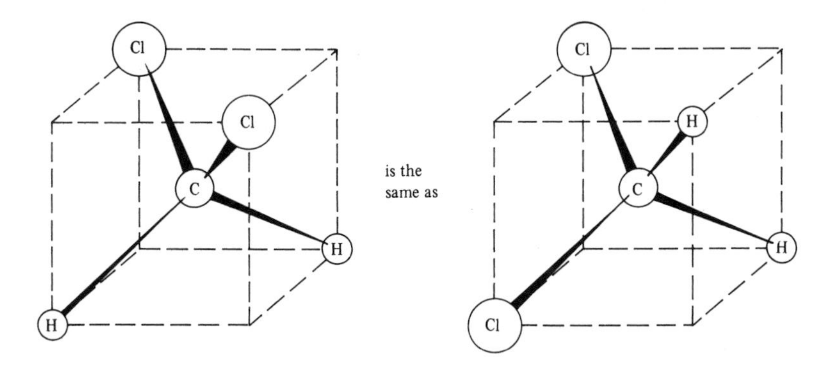

Fig. 1.3. There is only one dichloromethane molecule

Ionic and covalent bonds

An ionic bond is one due to the transfer of one or more electrons from one atom or group of atoms to another atom or group.

$$\begin{array}{cccc} \text{Na atom} + \text{Cl atom} & \rightarrow & \text{Na}^+ \text{ ion} + \text{Cl}^- \text{ ion} \\ (2,8,1) \quad (2,8,7) & & (2,8) \quad (2,8,8) \end{array}$$

One electron is transferred from the sodium atom to the chlorine atom, and the atoms thus become ions. The opposite charges of the ions attract each other. Electrons in the inner orbitals, i.e. the 2,8 of sodium and of chlorine take no part in forming bonds. Examples of ions are NH_4^+, Ca^{2+}, Al^{3+}, Br^-, SO_4^{2-}, $Cr_2O_7^{2-}$.

A covalent bond is one due to the sharing of two electrons between atoms. Two hydrogen atoms combine by sharing their single electrons. The pair of shared electrons revolves around both atoms, which in effect have the stable helium duplet. A chlorine atom has 7 electrons in its outer orbital. Two chlorine atoms combine by sharing two electrons, one from each atom. Each atom then has 6 electrons of its own in the outer orbitals plus the shared pair of the covalent bond, which revolves around both atoms. Therefore each atom has a complete outer octet of electrons and the stable argon structure, 2,8,8.

Properties of ionic and covalent substances

Ionic e.g. ($Na^+ + Cl^-$)	*Covalent* e.g. most organic substances
Crystalline solids	Liquids and gases at ordinary temperatures
High melting and boiling points	Low melting and boiling points
Electrolytes	Non-electrolytes
Insoluble in organic solvents, e.g. ether, ethanol, benzene	Soluble in organic solvents

If a covalent bond is between two identical atoms, e.g. C—C or H—H, the two electrons of the bond are shared equally and the bond is 100 per cent covalent. If the atoms are not the same the two electrons are not shared equally. One atom attracts the electron pair more than the other atom does, and therefore it has a small negative charge (and the other atom a small positive charge). Such a bond is described as *polar*. Each of the atoms, O, N, Cl and Br attract the electrons of a covalent bond more strongly than a carbon atom does, but a hydrogen atom attracts them slightly less strongly. The symbol δ is used to represent a partial charge arising from such *bond polarization*:

$$\overset{\delta+}{C}-\overset{\delta-}{O} \quad \overset{\delta+}{C}-\overset{\delta-}{N} \quad \overset{\delta+}{C}-\overset{\delta-}{Cl} \quad \overset{\delta+}{C}-\overset{\delta-}{Br} \quad \text{but} \quad \overset{\delta-}{C}-\overset{\delta+}{H}$$

Alternatively, an arrow is sometimes used:

$$C \mapsto Cl$$

Polar covalent compounds have properties with more ionic character than those of similar non-polar covalent compounds. For example, the melting and boiling points of CH_3Cl and C_6H_5Cl are higher than those of CH_4 and C_6H_6 respectively. Similarly the boiling point of ethanol, 78 °C, is much higher than those of hydrocarbons of similar molecular mass, e.g. C_3H_8, which boils at −42 °C, and also ethanol is much more soluble in water and much more reactive.

How covalent bonds break (bond fission)

In most chemical reactions involving organic compounds, a covalent bond between two atoms X and X or X and Y breaks and a new covalent bond between two atoms forms. In other words, the distribution of electrons around atomic nuclei changes. An understanding of the mechanism of reactions requires a knowledge of how bonds can break and why they do so.

A covalent bond between two atoms or groups may be represented by X:X or X:Y in which : denotes the shared electrons. The bond can break in two ways:

1. *Equal sharing of the electron pair* (*homolytic fission*). Each of the two atoms or groups takes one electron of the shared pair:

$$X{:}X \rightarrow X\cdot + \cdot X; \qquad X{:}Y \rightarrow X\cdot + \cdot Y$$

Each product, called a *free radical*, is an atom or group:

$$Cl{:}Cl \rightarrow Cl\cdot + \cdot Cl; \qquad H_3C{:}CH_3 \rightarrow H_3C\cdot + \cdot CH_3$$

A free radical has an unpaired electron and therefore is so reactive that it exists for only a fraction of a second and then reacts. Equal sharing is called homolysis or *homolytic fission*, and a reaction in which this occurs has a *free radical mechanism.* During homolytic bond fission the movement of a single electron can be depicted by a half-arrow.

$$Cl-Cl \rightarrow 2Cl\cdot$$

2. *Unequal sharing of the electron pair* (*heterolytic fission*). One atom or group takes both electrons of the shared pair:

$$X{:}X \rightarrow X^+ + {:}X^-; \qquad X{:}Y \rightarrow X^+ + {:}Y^- \text{ (or } X{:}^- + Y^+)$$

The products are ions. Unequal sharing is called heterolysis, and a reaction in which this occurs has a polar or 'ionic' mechanism. In *polar reactions* electron pairs are transferred from one species to another.

Each reagent $:X^-$ and $:Y^-$ possesses at least one pair of unshared electrons called a *lone pair*. The lone pair can be shared with a carbon or other nucleus, forming a new bond. Such reagents are *nucleophilic* ('nucleus loving') and are called *nucleophiles.* A nucleophile is a species which can donate an electron pair in a polar reaction. Common nucleophiles are H_2O, OH^-, Br^-, NH_3 and CN^-.

Each reagent X^+ and Y^+ has lost its share of an electron pair and can accept a share in a different lone pair. Reagants such as these are *electrophilic* ('electron loving') and are called *electrophiles.* An electrophile is a species which can accept an electron pair in a polar or 'ionic' reaction. Common electrophiles are H^+, Br^+ and NO_2^+.

Most organic reactions occur at carbon, nitrogen or oxygen atoms. A nucleophile seeks positions of positive charge and an electrophile seeks positions of high electron density. Examples are on pp. 33, 66, 80 and 94.

SUMMARY

Quantitative analysis

Carbon and hydrogen	Heat compound in oxygen. Absorb CO_2(g) in KOH(aq) and H_2O(g) in $CaCl_2$(s).
Nitrogen	Heat compound with excess CuO(s). Pass NO(g) and NO_2(g) over hot Cu and measure volume of N_2(g) formed.
Halogens	Heat compound with HNO_3 (fuming) and excess $AgNO_3$(s). Cool, add to water. Weigh dry AgCl, AgBr or AgI.
Sulphur	Heat compound with HNO_3 (fuming). Cool, dilute, add $Ba(NO_3)_2$(aq). Weigh dry $BaSO_4$.
Oxygen	100 minus the percentages of all other elements is the percentage of oxygen.

Physical methods of structure determination

Infrared absorption spectra. Each bond absorbs characteristic frequencies (wavelengths).

X-ray diffraction. Determines bond lengths and angles for solid crystals.

Electron diffraction by gas or vapour. Determines structure and bond lengths and angles.

Mass spectrometer. Determines relative molecular mass and structure.
High energy electrons form positive ions which give rise to peaks in the spectrum.

Bonds

An ionic bond is formed by transfer of electrons.

A covalent bond is formed by two electrons shared between atoms.

The four covalent bonds of the carbon atom in methane are arranged tetrahedrally, i.e. with maximum angle ($109\frac{1}{2}^\circ$) between them. Most covalent bonds are polar.

Homolytic fission is the breaking of a covalent bond to form two free radicals, e.g. $Cl\cdot$, $CH_3\cdot$.

Heterolytic fission is the breaking of a covalent bond to form two ions, e.g. $X{:}Y \rightarrow X{:}^- + Y^+$.

A nucleophilic reagent is electron-rich and has one pair of unshared electrons, e.g. OH^-, Br^-, CN^-, NH_3.

An electrophilic reagent is electron-deficient and seeks an unshared pair of electrons, e.g. H^+, Br^+, NO_2^+.

QUESTIONS

1. An organic compound contains sulphur. How would you determine the percentage by mass of sulphur in the compound? State the weighings that must be done.

2. Outline one method by which the percentage by mass of bromine in bromobenzene may be determined experimentally. Indicate the readings that are taken, and show how the percentage is calculated from the readings. (Br = 80; Ag = 108; C = 12; H = 1.)

3. Distinguish between the molecular formula and the empirical formulae of two named substances (the two formulae must be different for each compound selected).

4. 10 cm^3 of a gaseous alkane were mixed with 80 cm^3 (excess) of oxygen and exploded. After cooling to room temperature, the volume of gas remaining was 60 cm^3, and potassium hydroxide solution absorbed 30 cm^3. Calculate the formula of the alkane.

5. Calculate the volume of oxygen needed for the complete combustion of a mixture of 500 cm^3 of methane and 500 cm^3 of hydrogen. Show that the volume of the gaseous product is 500 cm^3. (All volumes are at room temperature and pressure.)

6. 1120 cm^3 (converted to s.t.p.) of a mixture of ethane and ethene was passed through a solution of bromine in ethanoic acid. 4 g of bromine was used up by the ethene and no bromine reacted with the ethane. Show that the percentage composition by volume of the gaseous mixture is 50 per cent.

7. 10 cm^3 of a gaseous hydrocarbon were exploded with 100 cm^3 (excess) of oxygen. 70 cm^3 of gas remained (after cooling to room temperature). Potassium hydroxide solution reduced this volume to 10 cm^3. Calculate the formula of the hydrocarbon.

8. 196 cm^3 of oxygen were added to 100 cm^3 of a mixture of hydrogen, methane and nitrogen. The mixture was now exploded and the volume after explosion was 165 cm^3. The residual gases were treated with aqueous potassium hydroxide and the final volume of gas was 131 cm^3. All volumes were measured at room temperature and pressure. Show that the hydrogen-methane-nitrogen mixture contained 42 cm^3 of hydrogen and 34 cm^3 of methane.

9. 0.27 g of a hydrocarbon were burnt in excess oxygen. The products were passed through calcium chloride tubes, which gained 0.27 g in mass, and through potassium hydroxide bulbs, which gained 0.88 g. Show that the empirical formula of the hydrocarbon is C_2H_3. If its relative molecular mass is 54, write its molecular formula and two possible displayed (graphic) formulae.

10. Write notes on the bonds between the atoms in (a) sodium chloride, (b) hydrogen, and (c) oxygen. State briefly how the nature of the bonds affects the properties of the compounds.

11. Explain what is meant by the statement that a C—C bond is non-polar but a C—Cl bond is polar. Why does a compound with polar bonds have a higher boiling point than a comparable compound with non-polar bonds, e.g. CH_4, 112 K; CH_3Cl, 249 K?

12. Write brief notes on infrared absorption spectra. Include notes on any two of the following: (a) a liquid-phase spectrum, e.g. hexane or hexene, (b) a comparison of the spectra of an alcohol and an isomeric ether, (c) a comparison of the spectra of a liquid aldehyde and ketone, and (d) the effect of strong hydrogen bonding on infrared spectra, e.g. those of carboxylic acids.

Chapter 2
ALKANES AND PETROLEUM

ORGANIC COMPOUNDS

The term inorganic is applied to minerals found on or in the earth's crust and to substances derived from them. In the eighteenth century the term organic was applied to fats, oils, sugars, proteins and other compounds containing carbon which were found in animals and plants. A special kind of *vital force* was supposed to be present in living things; organic compounds contained this vital force and could not be prepared except inside living things. In 1828 the organic compound carbamide (urea), $CO(NH_2)_2$, which is excreted by animals, was obtained by heating ammonium cyanate, NH_4CNO, a compound obtained from purely inorganic substances. Hundreds of organic compounds were soon prepared from inorganic substances, and the theory of vital force was therefore proved incorrect. Organic chemistry now means the *study of carbon compounds,* but the oxides and sulphides of carbon, carbonates, hydrogencarbonates and carbides are usually studied as part of inorganic chemistry. *Organic compounds* are those which contain carbon-carbon and/or carbon-hydrogen and/or carbon-halogen bonds.

One unique property of carbon is the ease with which its atoms join together. Many organic compounds have molecules which contain long stable chains of carbon atoms, for example, a molecule of natural rubber can contain 100 000 carbon atoms. A second unique property is that carbon atoms can form rings of three or more atoms, for example, one molecule of benzene or of a benzene compound contains a ring of six carbon atoms. Molecules of glucose and starch contain rings of five carbon atoms and one oxygen atom; a starch molecule consists of these rings linked together in a long chain and may contain as many as 18 000 carbon atoms. All living things contain proteins, and protein molecules have relative molecular masses between 5000 and 1 000 000 or so.

Most organic compounds can be classified into series; the compounds of one series are prepared by similar reactions and have similar properties. The first of these series we study is the alkanes, and the first three alkanes have the formulae CH_4, C_2H_6 and C_3H_8. There are 75 possible alkanes $C_{10}H_{22}$, and 366 319 with the formula $C_{20}H_{42}$. The advantage and importance of studying series of organic compounds rather than individual compounds are obvious.

ALKANES

Hydrocarbons contain only hydrogen and carbon. The alkanes, formerly called the *paraffins* (which means 'little affinity') are the first of four hydrocarbon series which we study. The general molecular formula is C_nH_{2n+2}, in which $n = 1$ or more. The carbon atoms in each molecule are in unbranched and branched chains and not rings. Compounds which contain such chains and no rings are described as *aliphatic* (from a Greek word meaning 'fat') because some are obtained from fats.

All names of alkanes end in *ane*, and the first part of each name indicates the number of carbon atoms in one molecule.

Name	*Formula*	*Relative molecular mass*	*M.p./K*	*B.p./K*	*Density/g cm^{-3}*
Methane	CH_4	16	90.7	111.7	0.424
Ethane	C_2H_6	30	89.9	184.5	0.546
Propane	C_3H_8	44	85.5	231.1	0.582
Butane	C_4H_{10}	58	134.8	272.7	0.579
Pentane	C_5H_{12}	72	143.4	309.2	0.626
Hexane	C_6H_{14}	86	177.8	341.9	0.659
Octane	C_8H_{18}	114	216.4	398.8	0.703
Decane	$C_{10}H_{22}$	142	243.5	447.3	0.730
Hexadecane	$C_{16}H_{34}$	226	291	550	0.780

(The densities are at 298 K for liquids and 6 K for alkanes which are gases at normal temperatures.)

Names of other alkanes include: heptane, nonane, undecane, dodecane, $C_{12}H_{26}$; tridecane, $C_{13}H_{28}$; tetradecane, $C_{14}H_{30}$; pentadecane, $C_{15}H_{32}$; heptadecane, $C_{17}H_{36}$; octadecane, $C_{18}H_{38}$.

The prefixes *meth, eth, prop* and *but* in the first four names are not systematic, but the prefixes from *pent* onwards are derived from Greek words, except *non* for 9 and *undec* for 11 which are derived from Latin. The names of the first four alkanes are so well established that it has been decided not to change them to systematic names.

Alkyl groups

An alkyl group is the part remaining if one hydrogen atom is removed from an alkane molecule. The general formula of alkyl groups is C_nH_{2n+1} and each group is univalent. The first four alkyl groups are methyl, CH_3-, ethyl, C_2H_5- or CH_3CH_2-, propyl, C_3H_7- or $CH_3CH_2CH_2-$, and butyl, C_4H_9- or $CH_3CH_2CH_2CH_2-$. The amine CH_3NH_2 is therefore called methylamine, and the halide C_2H_5Br is ethyl bromide (its systematic name is bromoethane).

Occurrence of alkanes

Methane occurs as *fire-damp* in gases released during the mining of coal and as *marsh-gas* in gases evolved from swamps, ponds and mud in which vegetable matter such as cellulose putrefies under water. *Natural gas* is found under the sea and over petroleum; it contains between 80 and 99 per cent of methane together with other gaseous alkanes. Methane is formed when coal is heated in the absence of air (destructive distillation of coal); coal gas contains about 30 per cent of methane by volume. *Petroleum* or crude oil contains as many as eighty alkanes; fractional distillation and refining of petroleum yields numerous products (p. 33). In some sewage works, the sewage is kept in large tanks at about 300 K until the bacteria present decompose substances in the sewage and yield a gas which consists of about 70 per cent methane and 30 per cent carbon dioxide. Most of the carbon dioxide is removed and the methane is then compressed and used as a motor fuel.

Laboratory preparation of alkanes

1. *From sodium salts of carboxylic acids.* Some organic acids (p. 175) contain the carboxyl group $-COOH$ and their sodium salts contain the group $-COONa$. A sodium salt is heated with barium hydroxide or sodalime, which is a mixture of anhydrous sodium hydroxide and calcium hydroxide. Sodalime is better than sodium hydroxide in this preparation because it is not deliquescent and it melts at a higher temperature.

$$\underset{\text{sodium ethanoate}}{2CH_3COONa(s)} + Ba(OH)_2(s) \rightarrow Na_2CO_3 + BaCO_3 + \underset{\text{methane}}{2CH_4(g)}$$

$$CH_3COONa(s) + NaOH(s) \rightarrow Na_2CO_3 + CH_4(g)$$

$$\underset{\text{sodium propanoate}}{C_2H_5COONa(s)} + NaOH(s) \rightarrow Na_2CO_3 + \underset{\text{ethane}}{C_2H_6(g)}$$

Methane. Sodium ethanoate crystals are hydrated, $CH_3COONa{\cdot}3H_2O$. Heat the crystals in an evaporating dish to drive off all the water of crystallization. Allow the anhydrous solid to cool and then powder it in a mortar. Add about twice the mass of sodalime or barium hydroxide to the mortar and grind the mixture to a fine powder. Add the mixture to a hard-glass boiling tube. Fit the tube with a stopper and delivery tube leading to a trough. Heat the tube fairly strongly until gas is evolved and collect the methane over water. Collect 8 jars or tubes of the gas and use in tests described on p. 20. Remove the delivery tube from the water before heating is stopped in order to prevent water sucking back into the hot tube. The apparatus is like that in Fig. 3.2, but without the glass wool.

The methane is not pure; it contains hydrogen and ethene, C_2H_4. The yield of ethane from sodium propanoate is only about 50 per cent because the product contains methane, ethene and hydrogen as impurities.

2. *Reduction of halogenoalkanes.* The reducing agent used is methanol or ethanol and an aluminium-mercury or zinc-copper couple. Prepare the aluminium-mercury couple by adding pieces of thin aluminium foil to mercury(II) chloride solution; mercury is deposited on the foil and the couple sinks in the solution. Prepare the zinc-copper couple by adding zinc foil to copper(II) sulphate solution; copper is deposited on the zinc. Dry the couple by washing with methanol.

$$2Al(s) + 3Hg^{2+}(aq) \rightarrow 2Al^{3+}(aq) + 3Hg(l)$$

$$Zn(s) + Cu^{2+}(aq) \rightarrow Zn^{2+}(aq) + Cu(s)$$

Mix dry methanol or ethanol with either dry iodomethane or dry iodoethane, which will form methane and ethane respectively. Place the mixture in a dropping funnel fitted into the neck of a flask containing the dry metal couple. Add the liquid slowly to the couple and collect the alkane over water. (Alcohols are weak acids (p. 133).)

$$CH_3I + 2[H] \text{ (from couple)} \rightarrow HI + \underset{\text{methane}}{CH_4}$$

$$C_2H_5I + 2[H] \text{ (from couple)} \rightarrow HI + \underset{\text{ethane}}{C_2H_6}$$

$$CH_3I + H^+(alc) + Zn \rightarrow I^- + Zn^{2+} + CH_4(g)$$

3. *Action of sodium on halogenoalkanes* (*Wurtz reaction*). Add a dry halogeno-alkane, or a mixture of two dry halogenoalkanes, to clean sodium and dry ethoxyethane (ether). The ether acts as a solvent for the halogen compounds. Reaction occurs spontaneously. Collect the gas over water.

$$CH_3I + 2Na + ICH_3 \rightarrow 2NaI + \underset{\text{ethane}}{CH_3CH_3}$$

$$C_2H_5I + 2Na + ICH_3 \rightarrow 2NaI + \underset{\text{propane}}{C_2H_5CH_3}$$

$$2C_2H_5Br(l) + 2Na(s) \rightarrow 2NaBr + C_2H_5C_2H_5 \text{ or } \underset{\text{butane}}{C_4H_{10}(g)}$$

The Wurtz reaction enables an alkane to be prepared from a halogen compound whose molecule contains only half the number of carbon atoms present in the alkane molecule. For example, hexane, $CH_3CH_2CH_2CH_2CH_2CH_3$, is prepared from bromopropane, $CH_3CH_2CH_2Br$.

Tests on methane or ethane

1. *Combustion.* (a) Ignite a jar of the alkane (which must be free from air). Note the luminosity of the flame. Observe if soot forms in the jar. Cover the jar as soon as the flame goes out. Test for the presence of carbon dioxide by adding calcium hydroxide solution (limewater) and shaking. (b) Ignite in a boiling-tube (not a jar) a mixture of air and alkane. Be prepared for an explosion. Note if moisture forms on the tube.

2. *Bromine.* Bromine is a dense dark-red liquid. Both the vapour and liquid are toxic, the vapour irritates the eyes and nose, and the liquid burns the skin. Experiments with liquid bromine must be done in a fume cupboard. (a) Use a teat pipette to add 4 drops of bromine to a gas-jar of air. The jar soon fills with the reddish-brown bromine vapour. Place this jar mouth to mouth over a jar of alkane and allow the bromine and alkane to mix. Note if the colour of the bromine changes and if any change is rapid or slow. Leave for 24 hours if necessary. (b) Add 2 drops of concentrated bromine water to a jar of alkane gas and shake well. Note if the bromine is decolorized and if any change is rapid or slow. If the bromine is decolorized, separate the jars and blow across their mouths; observe if misty fumes of hydrogen bromide are formed. (c) Repeat test (b) with a solution of bromine in tetrachloromethane, CCl_4.

3. *Potassium manganate(VII).* $KMnO_4$. Add about 2 cm^3 of very dilute potassium manganate(VII) solution to each of three test-tubes. Acidify one solution by adding about 2 cm^3 of dilute sulphuric acid, and make the second solution alkaline by adding sodium carbonate solution. Leave the third solution neutral.

Add the acidified, alkaline and neutral solutions separately to jars of the alkane, and shake well. Note if the pink solution is decolorized or changed and if any change is rapid or slow.

Since impurities in the methane or ethane may produce incorrect observations, repeat tests 2(b), 2(c) and 3 with pure hexane, a liquid alkane. Use about 10 drops of hexane in each reaction.

Homologous series

A series of compounds of similar structure in which each member differs from the next by the presence of one $-CH_2-$ group is called a homologous series. The members of the series are called *homologues.* The alkanes are the first homologous series in this book. Characteristics of a homologous series are:

1. All the homologues can be represented by a general formula.

2. The formula of one homologue differs from the formulae of the two homologues next to it, one on either side, by one more or one less $-CH_2-$ group, for example:

   CH_3CH_3 $\quad$ $CH_3CH_2CH_3$ $\quad$ $CH_3CH_2CH_2CH_3$

   and so on.

3. Homologues can usually be prepared by similar methods.

4. Homologues have similar chemical properties. Sometimes the chemical properties vary gradually with increasing relative molecular mass. Alkanes are so unreactive that it is difficult to demonstrate any variation in their reactions. However, higher alkanes (those with more carbon atoms per molecule) do not burn so readily as methane, ethane and lower alkanes.

5. The physical properties of homologues show a progressive change with increasing relative molecular mass, that is, as the number of carbon atoms per molecule increases. The melting points, boiling points and densities of alkanes rise with increasing relative molecular mass. Usually the solubility in water of homologues decreases with increasing relative molecular mass; the alcohols show this change clearly (p. 110).

6. Homologues contain at least one *functional group,* which is the particular group in the molecules most important in determining the properties. The functional group in alkanes is the unreactive group C—H, and the group in the halogenoalkanes is C—F, C—Cl, C—Br or C—I. The functional group in the hydrocarbons called the alkenes is C=C, which represents a double bond between two carbon atoms.

Various formulae of alkanes

The *empirical formula* of a compound expresses the composition by mass of the compound and it is the simplest formula which shows the ratio of the number of atoms in one molecule. The *molecular formula* shows the actual number of each kind of atom present in one molecule of the compound.

Compound	*Empirical formula*	*Molecular formula*
Methane	CH_4	CH_4
Butane	C_2H_5	C_4H_{10}
Ethyne	CH	C_2H_2
Benzene	CH	C_6H_6
Methanal	CH_2O	CH_2O
Glucose	CH_2O	$C_6H_{12}O_6$

The *structural formula* shows the sequence and arrangement of the atoms in a molecule, e.g. $CH_3CH_2CH_2CH_3$ for butane, and CH_3CH_2OH for ethanol. A *displayed formula* (or *graphic formula*) shows the spatial arrangement of the atoms or groups of atoms and their linkages, represented by a dash (–), projected on to a plane.

Alkane	*Structural formula*	*Displayed (graphic) formula*
Methane	CH_4	$\begin{array}{c} H \\ | \\ H—C—H \\ | \\ H \end{array}$ or $\begin{array}{ccccc} & & H & & \\ & & | & & \\ & / & C & \backslash & \\ H & & | & & H \\ & & H & & \end{array}$
Ethane	CH_3CH_3	$\begin{array}{c} H\ H \\ |\ \ | \\ H—C—C—H \\ |\ \ | \\ H\ H \end{array}$
Propane	$CH_3CH_2CH_3$	$\begin{array}{c} H\ H\ H \\ |\ \ |\ \ | \\ H—C—C—C—H \\ |\ \ |\ \ | \\ H\ H\ H \end{array}$ or $\begin{array}{ccccc} & & H\ H & & \\ & & \backslash / & & \\ H & & C & & H \\ \backslash & / & & \backslash & / \\ C & & & & C \\ / \backslash & & & & / \backslash \\ H\ H & & & & H\ H \end{array}$

In the above formulae each quadrivalent carbon atom is joined by four covalent bonds to four other atoms, and each univalent hydrogen atom is joined by one covalent bond to a carbon atom. The displayed formulae have the disadvantage of representing three-dimensional molecules in only two dimensions. Make ball-and-spring and space-filling models of the above molecules.

Isomerism of alkanes

Four carbon atoms can be joined in two ways: an unbranched chain of four atoms or a branched chain having the fourth atom joined to the middle of three other atoms:

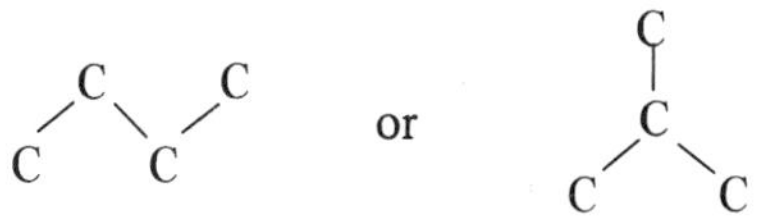

```
   C   C              C
  / \ /               |
 C   C       or       C
                     / \
                    C   C
```

Two alkanes have these arrangements of carbon atoms in their molecules. The molecular formula of both alkanes is C_4H_{10}, and their names, structural and displayed formulae are:

$CH_3CH_2CH_2CH_3$ $\quad\quad$ $CH_3CH(CH_3)CH_3$ or $(CH_3)_2CHCH_3$

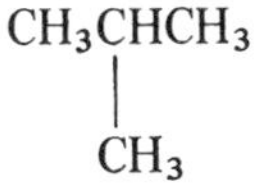

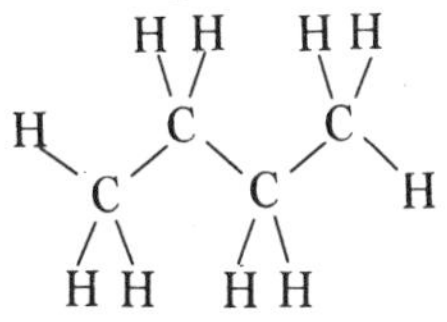

```
     H H   H H               H H   H H
      \/    \/                \/    \/
 H     C     C               C     C
  \  /  \  /  \           /   \   /  \ H
   C     C     H        H     C        
  /\    /\                  /  \ C—H
 H H   H H                 H    / \
                               H   H
     butane                methylpropane
```

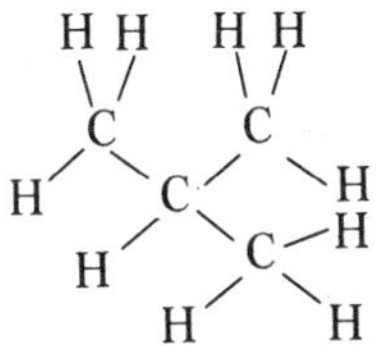

Compounds which have the same molecular formulae but different displayed formulae are called *isomers* or *isomeric compounds*, and the existence of such compounds is called *isomerism*. There are five isomeric alkanes C_6H_{14} and nine isomers C_7H_{16}.

The hydrocarbon name of a branched chain compound is derived from the longest unbranched or continuous chain of carbon atoms in the molecule. The longest chain in the formula on the right above has three carbon atoms and the compound is therefore a propane derivative. The CH_3– or methyl group is attached to carbon atom number 2, the middle of the three atoms, and therefore the compound is called 2-methylpropane.

Three alkanes are isomers with the formula C_5H_{12}. The longest chain can have either five, four or three carbon atoms:

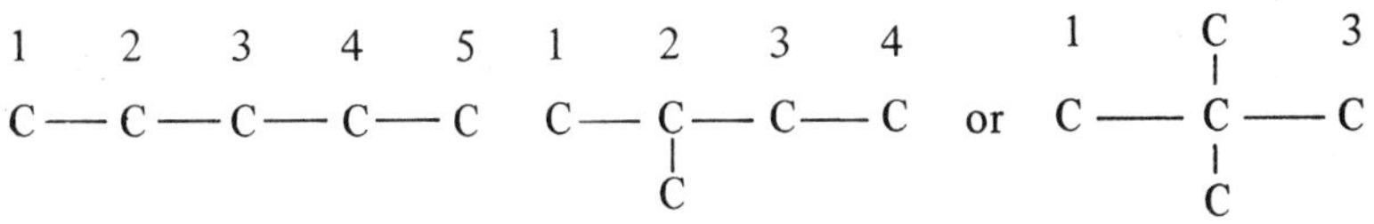

```
1   2   3   4   5    1   2   3   4        1    C    3
                                               |
C — C — C — C — C    C — C — C — C   or   C —  C  — C
                         |                     |
                         C                     C
```

The names, structural and displayed formulae of the isomers are:

$CH_3CH_2CH_2CH_2CH_3$ | $(CH_3)_2CHCH_2CH_3$ or $CH_3CH_2CH(CH_3)_2$ | $(CH_3)_4C$ or $CH_3C(CH_3)_2CH_3$

pentane | 2-methylbutane | 2,2-dimethylpropane

The longest carbon chain is numbered in such a way that attached groups have the lowest possible numbers. The third isomer has two methyl groups attached to the second carbon atom of the propane chain, and their positions are represented by two figures separated by a comma, that is, by 2,2-.

If the alkyl groups in one molecule are different, they are numbered in alphabetical order, for example, ethyl comes before methyl:

$$\begin{array}{l} \text{6 5 4 3 2 1} \\ CH_3CH_2\underset{\underset{\displaystyle CH_3}{|}}{\underset{|}{CH_2}}\!\!\!\!\!\!\!\!CHCH_2\underset{\underset{\displaystyle CH_3}{|}}{CH}CH_3 \end{array}$$

$CH_3CH_2CH(C_2H_5)CH_2CH(CH_3)CH_3$ is 4-ethyl-2-methylhexane, and not 2-methyl-4-ethylhexane

This compound is not 3-ethyl-5-methylhexane (obtained by numbering the hexane chain from left to right) because the numbers must be the lowest possible, and these are 2 and 4 if the hexane chain is numbered from the right.

Formulae of alkyl groups

$$H-\overset{\displaystyle H}{\underset{\displaystyle H}{C}}- \quad \text{or } CH_3-$$

methyl

$$H-\overset{\displaystyle H}{\underset{\displaystyle H}{C}}-\overset{\displaystyle H}{\underset{\displaystyle H}{C}}-\overset{\displaystyle H}{\underset{\displaystyle H}{C}}- \quad \text{or } C_3H_7- \text{ or } CH_3CH_2CH_2-$$

propyl

$$H-\overset{\displaystyle H}{\underset{\displaystyle H}{C}}-\overset{\displaystyle H}{\underset{\displaystyle H}{C}}- \quad \text{or } C_2H_5- \text{ or } CH_3CH_2-$$

ethyl

$$H-\overset{\displaystyle H}{\underset{\displaystyle H}{C}}-\overset{\displaystyle H}{\underset{|}{C}}-\overset{\displaystyle H}{\underset{\displaystyle H}{C}}-H \quad \text{or } (CH_3)_2CH-$$

1-methylethyl

Properties of methane and ethane

The properties of other alkanes are similar to those of methane and ethane, but differences are mentioned when they exist.

Physical properties. Methane, ethane, propane, butane and 2-methylpropane are colourless, odourless, non-poisonous gases. The alkanes from pentane to heptadecane are liquids under ordinary conditions; octadecane, $C_{18}H_{38}$, and higher alkanes are solids. Each extra CH_2 group increases the boiling point of the alkane, but each increase is smaller than the previous one. The densities show a gradual change up the series. Alkanes are only slightly soluble in water and are more soluble in ethanol. Liquid alkanes are miscible with one another.

The melting points of alkanes do not rise regularly. However, melting points of alkanes with an odd number of carbon atoms rise regularly, and melting points of alkanes with an even number of carbon atoms also rise regularly.

Chemical properties. Alkane molecules contain only single bonds (C—H and C—C) between the atoms. Alkanes are therefore examples of *saturated* compounds. Atoms cannot be added to an alkane molecule because it has no spare valency bonds. An atom or group can enter an alkane molecule only by replacing one or more hydrogen atoms. A reaction in which one atom or group is replaced by another one is called a *substitution reaction.*

$$CH_4 + XY \rightarrow CH_3X + HY$$

$$CH_4 + X_2 \rightarrow CH_3X + HX$$

Alkanes are inert and do not react with common reagents such as acids, alkalis and oxidizing agents.

1. *Thermal decomposition.* Methane and ethane are stable to heat. Propane and higher alkanes decompose at high temperatures and form two or more simpler substances. The changes are called *cracking* because one molecule 'cracks' to form two or more hydrocarbon molecules (and sometimes hydrogen) with fewer carbon atoms per molecule than the original alkane. Chain breaking and loss of hydrogen (*dehydrogenation*) occur during cracking of alkanes:

$$C_3H_8 \rightarrow CH_4 + \underset{\text{ethene}}{C_2H_4} \quad \text{(chain breaking)}$$

$$C_3H_8 \rightarrow H_2 + \underset{\text{propene}}{C_3H_6} \quad \text{(dehydrogenation)}$$

$$C_{10}H_{22} \rightarrow C_3H_6 + \underset{\text{heptane}}{C_7H_{16}} \quad \text{(chain breaking)}$$

2. *Combustion.* Gaseous and volatile alkanes such as petrol and kerosine form explosive mixtures with air or oxygen. Alkanes burn in a plentiful supply of air or oxygen to form carbon dioxide and water:

$$CH_4 + 2O_2 \rightarrow CO_2 + 2H_2O$$

$$2C_2H_6 + 7O_2 \rightarrow 4CO_2 + 6H_2O$$

Methane burns with a blue, faintly luminous flame similar to the non-luminous flame of a bunsen burner using natural gas; ethane and propane burn with bright flames. The flames become more luminous as the number of carbon atoms in the alkane molecules increases because the percentage by mass of carbon also increases. This percentage is 75 in methane, 80 in ethane, 81.8 in propane, and 82.7 in butane. The ease with which alkanes burn explains their use as fuels; petrol, kerosine, diesel oil, fuel oil, 'bottled gas' and natural gas are all mixtures of alkanes. Volatile alkanes form explosive mixtures with air, and this property is used in engines which burn petrol, kerosine and diesel oil.

Some carbon monoxide and carbon are formed when alkanes burn in a limited supply of air:

$$CH_4 + O_2 \rightarrow C(s) + 2H_2O$$

$$2CH_4 + 3O_2 \rightarrow 2CO(g) + 4H_2O$$

The incomplete combustion of alkanes in internal combustion engines results in the presence of poisonous carbon monoxide in the exhaust gases. Carbon is deposited as soot on the pistons and valves of the engines and makes them inefficient. An explosive gas called *fire-damp* present in coal mines is mainly methane; the poisonous gas known as *after-damp* formed by an explosion in a coal mine is carbon monoxide.

A mixture of chlorine gas and methane or ethane explodes either when exposed to bright sunlight or ultraviolet light or when sparked:

$$CH_4 + 2Cl_2 \rightarrow C(s) + 4HCl$$

$$C_2H_6 + 3Cl_2 \rightarrow 2C(s) + 6HCl$$

3. *Halogens.* Alkanes and chlorine react in diffused sunlight. Substitution reactions occur in which hydrogen atoms are replaced successively by chlorine atoms:

$$CH_4 + Cl_2 \rightarrow HCl + \underset{\text{chloromethane}}{CH_3Cl}$$

$$CH_3Cl + Cl_2 \rightarrow HCl + \underset{\text{dichloromethane}}{CH_2Cl_2}$$

$$CH_2Cl_2 + Cl_2 \rightarrow HCl + \underset{\text{trichloromethane}}{CHCl_3}$$

$$CHCl_3 + Cl_2 \rightarrow HCl + \underset{\text{tetrachloromethane}}{CCl_4}$$

Six chloroethanes are formed when ethane and chlorine react in diffused sunlight:

C_2H_5Cl, $C_2H_4Cl_2$, $C_2H_3Cl_3$, $C_2H_2Cl_4$, C_2HCl_5 and C_2Cl_6

Liquid bromine reacts less vigorously than chlorine. Bromine, bromine water and bromine dissolved in tetrachloromethane are each decolorized by excess alkanes, especially in the presence of bright sunlight. Reaction is very slow in the dark.

$$CH_4 + Br_2 \rightarrow HBr + \underset{\text{bromomethane}}{CH_3Br}$$

$$C_6H_{14} + Br_2 \rightarrow HBr + \underset{\text{bromohexane}}{C_6H_{13}Br}$$

Further substitution products formed by methane are CH_2Br_2, $CHBr_3$ and CBr_4. Ethane forms six bromoethanes.

Iodine does not react with alkanes. The order of reactivity with alkanes of the halogens is $Cl > Br > I$.

4. *Potassium manganate*(*VII*), $KMnO_4$, does not react with alkanes.

5. *Concentrated sulphuric acid* has no action on alkanes, but it may char impurities in them and cause a darkening in colour.

6. *Nitration.* Concentrated nitric acid does not react with alkanes. Methane and ethane can be nitrated with nitric acid vapour (a process called *vapour phase nitration*) at about 720 K and 10 atmospheres pressure:

$$C_2H_6(g) + HNO_3(g) \rightarrow H_2O + \underset{\text{nitroethane}}{C_2H_5NO_2}$$

Nitroalkanes (and chloroalkanes) are used as industrial solvents.

Van der Waals' forces between alkane molecules

In a hydrogen chloride molecule, H—Cl, the electron pair of the bond is not shared equally because the chlorine nucleus attracts the two electrons more than the hydrogen nucleus does (p. 10). As a result the chlorine atom has a small negative charge and the hydrogen atom has a small positive charge and the two charges together constitute a *dipole*. Thus hydrogen chloride molecules are *polar* and have a *permanent electrical dipole*:

$$\overset{\delta+}{H}-\overset{\delta-}{Cl} \text{ or } H \rightarrow\!\!\!\!\!+ Cl$$

The only bonds in alkane molecules are C—C and C—H, and the two electrons of each bond are shared almost equally by the atoms. Thus each bond is almost symmetrical, and in any alkane molecule the centre of positive charge of the protons and the centre of negative charge of the electrons practically coincide.

Alkane molecules are almost non-polar but do have a very weak dipole:

$$\overset{\delta-}{C}-\overset{\delta+}{H}$$

The forces between molecules in the liquid or solid state are called van der Waals' forces. They exist even between the molecules (which are also atoms) of helium and other noble gases; otherwise the gases would not liquefy or solidify.

Alkane molecules tend to arrange themselves in the liquid and solid states so that the dipole of one molecule attracts the dipole of a neighbouring molecule; a partial positive charge is next to a partial negative charge and there is an attraction between the two. In addition to or instead of a permanent dipole a molecule can have a *temporary dipole.* At a particular moment the electron charge cloud may be slightly displaced from its normal position, so that the centres of positive and negative charge no longer coincide. The molecule is thereby temporarily polarized, and this polar molecule may then induce a temporary dipole in a neighbouring molecule. The interaction between the two temporary dipoles causes attraction between the molecules. Permanent and temporary dipoles therefore explain van der Waals' forces.

Boiling points of alkanes

The boiling points of unbranched-chain alkanes rise as the number of carbon atoms per molecule increases. For example: methane, −162 °C; ethane, −88 °C; propane, −42 °C; butane, 0 °C. These four alkanes are all gases at room temperature, which shows that the forces of attraction between them are very weak. A larger alkane molecule has more electrons and a larger electron cloud than a smaller molecule and is therefore more easily distorted by a temporary dipole. Therefore van der Waals' forces between larger molecules are greater, it is more difficult to transfer the molecules from the liquid to the gaseous state, and the boiling points are higher.

The boiling points of alkane isomers are different. For example, those of the three C_5H_{12} isomers are:

Pentane 36 °C	2-Methylbutane 28 °C	2,2-Dimethylpropane 10 °C
$CH_3CH_2CH_2CH_2CH_3$	$(CH_3)_2CHCH_2CH_3$	$(CH_3)_4C$

The long pentane molecule has the largest surface area and the 2,2-isomer the smallest because it is almost a sphere. A pentane molecule has more neighbouring molecules and therefore the number of dipole-dipole attractions and also temporary-dipole–induced-dipole attractions (i.e. van der Waals' forces) is greatest; the forces are clearly smallest in the 2,2-isomer. Therefore pentane has the highest boiling point because the greatest energy is required to overcome the forces of attraction, and the 2,2-isomer has the lowest boiling point.

Melting points of alkanes

Some melting points are:

Methane, −182 °C   Ethane, −183 °C

Propane, C_3H_8, −188 °C   Butane, C_4H_{10}, −138 °C

Pentane, C_5H_{12}, −130 °C   Hexane, C_6H_{14}, − 95 °C

Heptane, C_7H_{16}, −90 °C   Octane, C_8H_{18}, −57 °C

Those of methane and ethane seem anomalous because they are very close to and not much lower than that of propane. A methane molecule has only one carbon atom and an ethane molecule two, which cannot be in a zig-zag chain. The carbon atoms in propane and all higher alkanes are arranged in a zig-zag chain. Methane and ethane molecules can pack together more closely than zig-zag chain molecules (space-filling models demonstrate the truth of this) and van der Waals' forces are greater, causing higher melting points.

Alkanes with an even number of carbon atoms per molecule pack more closely in solids than alkanes with odd numbers of carbon atoms (demonstrate this with models). This explains the relatively higher melting points of the C_4, C_6, etc. alkanes compared with the C_3, C_5, etc. alkanes. Plot two graphs of melting points against molecular mass, one for odd-numbered and another for even-numbered alkanes. They are two smooth curves. One graph for all alkanes is a zig-zag line.

The melting points rise for the unbranched-chain alkanes if those of methane and ethane are omitted. The rise is due to the greater temporary-dipole–induced-dipole attractions (van der Waals' forces) between molecules. The larger the molecule, the greater the temporary dipoles, the stronger the forces of attraction between the molecules, and therefore the higher the melting points.

The melting points of alkane isomers are different. For example those of the three C_5H_{12} isomers are: pentane, −130 °C; 2-methylbutane, −160 °C; 2,2-dimethylpropane, −16 °C. The melting points depend on how closely the molecules can pack together in the solid state. The compact, almost spherical, 2,2-isomer molecules pack most closely, have stronger van der Waals' forces, and therefore the highest melting point. However, the one branch on the chain of the 2-isomer makes packing of the molecules the most difficult and causes the lowest melting point. Even the unbranched-chain molecules of pentane do not pack well, and its melting point is low but between those of the other two isomers.

Uses of alkanes

Alkanes are more widely used than any other organic compounds.

Natural gas is mainly methane, and is used for heating and lighting. *Bottled gas, Calor gas, Afrigas and liquefied petrol gas (lpg)* are mixtures of propane, butane, and methylpropane. These alkanes are stored as liquids under pressure in steel cylinders; they become gases when the pressure is released. They are used for heat, power and light in houses, caravans, workshops, and so on; the cylinders are portable and do not require fixed gas pipes. Bottled gas is sometimes used instead of ethyne for cutting and welding metals because its flame is very hot. Propane and butane are used as bulk fuels in industry because they are easy to handle and free from sulphur compounds, which would burn to produce corrosive sulphur dioxide.

Petrol, kerosine, diesel oil and *fuel oil* are mixtures of alkanes obtained from petroleum (p. 33). Petrol usually contains alkanes from C_5H_{12} to C_9H_{20}, which boil below 420 K. Kerosine usually contains alkanes from $C_{10}H_{22}$ to $C_{15}H_{32}$, with boiling points between 420 and 520 K. Diesel oil and fuel oil are mixtures of higher alkanes, which are less volatile and boil between 520 and 670 K. Various higher alkanes are used in the manufacture of soaps, fats, detergents, plastics, lacquers, and varnishes.

Chloroalkanes. Chloromethane and dichloromethane are prepared commercially by thermal chlorination of methane at 670 K in the dark. Ethane is also chlorinated at 670 K to produce chloroethane. Uses of these halogen compounds are on p. 99.

Carbon black. Natural gas is burnt in a limited supply of air:

$$CH_4 + O_2 \rightarrow C(s) + 2H_2O$$

The carbon is used to manufacture printing inks, shoe polishes, gramophone records and rubber tyres. It makes the tyres stronger.

Hydrogen. Natural gas is oxidized by steam in the presence of a nickel catalyst or by oxygen:

$$CH_4 + H_2O \xrightarrow{Ni} CO + 3H_2$$

$$2CH_4 + O_2 \longrightarrow 2CO + 4H_2$$

The mixture of carbon monoxide and hydrogen is used as a fuel. It is also used as a source of hydrogen by first oxidizing the monoxide with steam in the presence of an iron catalyst at 800 K:

$$CO + H_2O \xrightarrow{Fe} CO_2 + H_2$$

The carbon dioxide is removed by dissolving it in potassium carbonate solution or in water under pressure. The hydrogen is used in the Haber process for the synthesis of ammonia.

Ethyne. Natural gas is heated to about 1800 K for about 0.01 s and the gases produced are cooled rapidly to reduce decomposition of the ethyne:

$$2CH_4 \xrightarrow{1800\,K} C_2H_2 + 3H_2 \text{ (dehydrogenation)}$$

Methanol. Naphtha is a mixture of light hydrocarbons obtained by fractionation of petroleum. Naphtha and steam are passed over a nickel catalyst, and a mixture of carbon monoxide and hydrogen is formed. This mixture is then passed over a catalyst at about 600 K and under a pressure of 300 atm.

$$2H_2 + CO \xrightarrow{\text{ZnO catalyst}} \underset{\text{methanol}}{CH_3OH}$$

Other products manufactured from petroleum include plastics, synthetic fibres, synthetic rubber, detergents, insecticides, weedkillers, and industrial solvents (pp. 52, 65, 80 and 217).

Halogenation of alkanes (homolytic reactions)

Methane and chlorine react in diffused sunlight

$$CH_4(g) + Cl_2(g) \rightarrow HCl(g) + \underset{\text{chloromethane}}{CH_3Cl(g)}$$

The reaction has a free radical mechanism and the free radical is a chlorine atom. A chlorine molecule absorbs light and the light energy causes homolytic fission:

$$Cl{:}Cl \rightarrow Cl\cdot + \cdot Cl \; (\textit{initiating reaction})$$

A series of two *propagating reactions* follows this initiating reaction. The whole process is a *chain reaction* because one chlorine atom causes a series of successive reactions, and these lead to the production of thousands of molecules of the products. The two propagating reactions are:

$$Cl\cdot + CH_4 \rightarrow HCl + H_3C\cdot, \quad \text{then} \quad H_3C\cdot + Cl_2 \rightarrow H_3CCl + Cl\cdot, \quad \text{etc.}$$

$H_3C\cdot$ is a methyl radical. Each dot in the above equations represents an unpaired electron. Full displayed formulae may be used in the last two equations above:

$$\begin{array}{c}H\\|\\H{-}C{-}H\\|\\H\end{array} + \cdot Cl \rightarrow \begin{array}{c}H\\|\\H{-}C\cdot\\|\\H\end{array} + H{-}Cl$$

$$\begin{array}{c}H\\|\\H{-}C\cdot\\|\\H\end{array} + Cl{-}Cl \rightarrow \begin{array}{c}H\\|\\H{-}C{-}Cl\\|\\H\end{array} + \cdot Cl$$

The propagating reactions continue until they are ended by one or more of the following *limiting* or *terminating reactions*:

$$Cl\cdot + \cdot Cl \rightarrow Cl_2$$

$$H_3C\cdot + \cdot CH_3 \rightarrow \underset{\text{ethane}}{C_2H_6}$$

$$H_3C\cdot + \cdot Cl \rightarrow CH_3Cl$$

Further chlorination produces dichloromethane. The monochloro-compound reacts with a chlorine atom to produce a free radical which then reacts with a chlorine molecule:

$$Cl\cdot + CH_3Cl \rightarrow ClH_2C\cdot + HCl$$

$$ClH_2C\cdot + Cl_2 \rightarrow CH_2Cl_2 + Cl\cdot$$

Displayed formulae make these reactions clearer:

$$Cl-\underset{H}{\overset{H}{\underset{|}{\overset{|}{C}}}}-H + \cdot Cl \rightarrow Cl-\underset{H}{\overset{H}{\underset{|}{\overset{|}{C}}}}\cdot + HCl$$

$$Cl-\underset{H}{\overset{H}{\underset{|}{\overset{|}{C}}}}\cdot + Cl-Cl \rightarrow Cl-\underset{H}{\overset{H}{\underset{|}{\overset{|}{C}}}}-Cl + \cdot Cl$$

Further chlorination produces trichloromethane and tetrachloromethane by similar mechanisms.

The action of bromine on methane is similar to that of chlorine. The chain reactions are much shorter and the reaction is therefore much slower. Iodine and methane do not react.

PETROLEUM

Petroleum (crude oil)

Formation. Petroleum was probably formed millions of years ago in shallow seas that contained plenty of animal and vegetable living organisms. The dead organisms settled on the sea-bed, were gradually covered with thick layers of sand and clay, and decomposed under great pressures and at high temperatures to form oil. The oil usually occurs in porous rock such as sandstone and has natural gas above it.

Composition. Petroleum is a black viscous liquid and sometimes it appears greenish. It is a mixture of many hydrocarbons and most are alkanes with from five to forty or more carbon atoms per molecule. The natural gas above the oil is mainly methane, but it may contain up to 20 per cent of other gaseous alkanes such as ethane, propane and butane. Holes are drilled down through the rocks to the petroleum, which is forced out by the pressure of the natural gas. Later the remaining petroleum must be pumped out.

Refining of petroleum

Petroleum is separated by fractional distillation (fractionation) into various mixtures or fractions, each of which consists of hydrocarbons of similar composition and boils within a limited temperature range. Since the composition of petroleum varies greatly even when obtained from one region, the fractions and their temperature ranges also vary. The products of petroleum refining supply a large and increasing proportion of the world's heat, light and power, and also supply more than half of its chemicals.

Fraction	*Approximate boiling range/K*	*Uses*
Hydrocarbon gases or petroleum gases	Below 310	Bottled gas for fuel
Gasoline (petrol)	310 to 420	Petrol, aviation fuel
Kerosine or naphtha	400 to 500	Paraffin oil for heat, light and fuel Jet fuel. Rocket fuel (first stage) Cracked to form petrol and alkenes
Gas oil Diesel oil	500 to 600	Cracked to form petrol and alkenes Fuel for diesel engines
Heavy oil Fuel oil	600 to 700	Lubrication Fuel for furnaces
Waxes	About 700	Vaseline, greases, candles, polishes
Bitumen (pitch)	About 700	Road surfaces and aircraft runways Waterproofing and protection of roofs and underground pipes

Hydrocarbon gases comprise methane, ethane, propane, butane and methylpropane. Some are used to heat the refinery furnace, and the rest is sold as bottled gas.

Gasoline (petrol) and kerosine (paraffin oil) are colourless liquid alkanes and gasoline is the more volatile. Gasoline is the most important fraction in the USA.

Naphtha is as important as gasoline in Europe. It is a mixture of light hydrocarbons, which are unbranched and branched chain alkanes, and some benzene compounds. It is cracked to form petrol, fuel oil, and alkenes such as ethene, propene and butene (p. 35).

Bitumen is the black residue of the refining process. It is a shiny black brittle solid or a black viscous liquid, depending on the crude oil and the refinery.

Fractional distillation or refining of petroleum

Liquid petroleum is pumped through pipes heated to about 700 K in a furnace; hydrocarbon gases formed in the refining process are burnt to heat the furnace. The petroleum vapours pass from the furnace into a tower about 60 m high in which the temperature decreases from 700 K at the bottom to 310 K at the top. Inside the tower are sets of trays on which condensed fractions collect. A bubble cap is over each tray so that vapour passing up the tower must bubble through liquid in the trays. The less volatile constituents of the vapour condense in the trays and the more volatile constituents of the liquid vaporize and pass up the tower. Excess liquid on each tray passes down an overflow pipe to the warmer tray below and is distilled again.

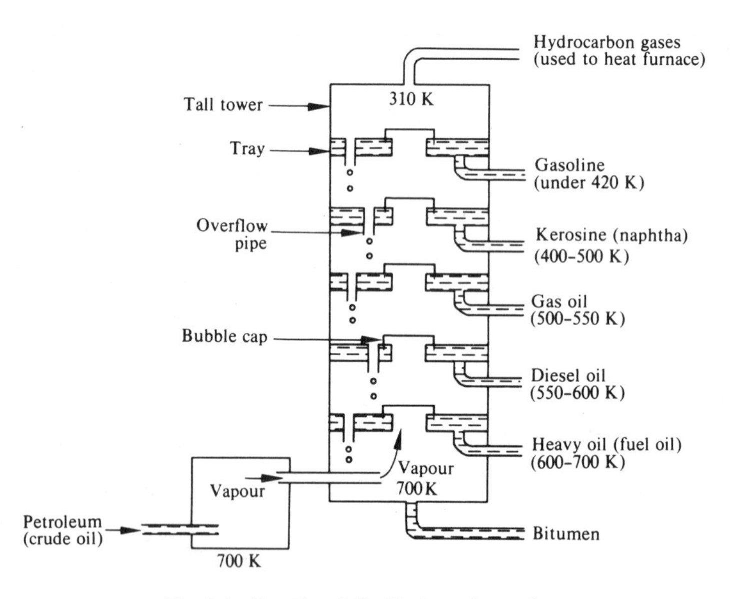

Fig. 2.1. Fractional distillation of petroleum

Hydrocarbon gases escape at the top of the tower. Fractions of liquid run off from the trays are treated by various processes (distillation, crystallization, solvent extraction, absorption on charcoal, and so on) to obtain pure products.

Sulphur from petroleum and natural gas

Natural gas and hydrocarbon gases obtained by refining petroleum contain hydrogen sulphide, which is absorbed in special solvents. The hydrogen sulphide is removed from the solvents and about one-third is burnt in air to produce sulphur dioxide, which is used to oxidize the remaining two-thirds to sulphur:

$$2H_2S + 3O_2 \rightarrow 2H_2O + 2SO_2$$

$$2H_2S + SO_2 \rightarrow 2H_2O + 3S(s)$$

This method is the world's third largest source of sulphur.

Commercial cracking of oils

Cracking is the process of breaking down hydrocarbons of high molecular mass into ones of smaller mass with formation of alkenes (or aromatic hydrocarbons):

$$\underset{\text{(large molecule)}}{\text{Alkane}} \rightarrow \underset{\text{(small molecule)}}{\text{Alkane}} + \text{Alkene (+ Hydrogen)}$$

Four chemical changes which occur during the cracking (p. 25) or reforming (p. 71) of hydrocarbon molecules are: the breaking of carbon chains, dehydrogenation, conversion of unbranched-chain molecules into branched-chain molecules, and conversion of chains of carbon atoms into rings of atoms (including benzenoid compounds).

Most petrol is a mixture of alkanes whose molecules have between 5 and 9 carbon atoms. Fractional distillation of petroleum does not supply enough petrol to meet the large demand. Kerosine, naphtha and gas oil are cracked or reformed in order to produce more petrol. Cracking also produces alkenes such as ethene by dehydrogenation:

$$-CH_2-CH_2- \rightarrow H_2 + \underset{\text{an alkene}}{-CH{=}CH-}$$

Thermal cracking. Gas oil is heated at 800 K under pressure so that it remains a liquid. The product is then fractionated to obtain petrol and other useful hydrocarbons.

Steam cracking. Naphtha vapour is mixed with steam, an inert diluent, and heated to 1200 K for less than a second. The naphtha cracks to form 30 per cent petrol and fuel oil, 50 per cent ethene and other alkenes, and some methane and hydrogen. Benzene compounds in the naphtha are not cracked.

Catalytic cracking. The catalyst used in an aluminosilicate of about 10 per cent aluminium oxide, Al_2O_3, spread through porous natural or synthetic silicon(IV) oxide, SiO_2. The temperature and pressure are much lower than in thermal cracking, and the products are different. The catalysts cause the formation of benzene hydrocarbons (p. 70) and of branched chain hydrocarbons from unbranched-chain compounds:

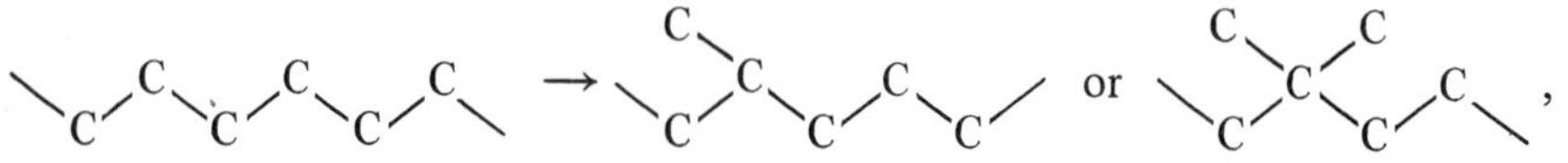

and so on.

Petrol which contains some of these hydrocarbons burns more efficiently in modern engines than pure unbranched chain alkanes.

A mixture of petrol vapour and air enters each cylinder of an engine as its piston moves down. The mixture is compressed as the piston moves up and is ignited by a spark when the piston is at the top of its movement or 'stroke'. However, petrol which consists largely of unbranched-chain alkanes tends to explode even before the spark passes; the premature explosions damage the piston and reduce the efficiency of the engine. The engine is said to be 'pinking' or 'knocking'. Petrol that contains benzene, ethanol, methanol and branched-chain alkanes does not cause 'knocking'.

SUMMARY

Alkanes

C_nH_{2n+2}. Functional group: C—H.

Preparation

1. Heat sodium carboxylate with sodalime.

$$CH_3COONa + NaOH \rightarrow \underset{\text{methane}}{CH_4}$$

2. Halogenoalkane and Al/Hg or Zn/Cu couple.

$$C_2H_5I(alc) \rightarrow \underset{\text{ethane}}{C_2H_6}$$

3. Halogenoalkane (ether) and dry sodium.

$$CH_3I(ether) + Na \rightarrow \underset{\text{ethane}}{CH_3CH_3}$$

Reactions

Heat	'Cracking' of larger alkane molecules Chain breaking and dehydrogenation
Combustion	$CO_2 + H_2O$ in air. Methane has blue, faintly luminous flame. C + HCl with chlorine in sunlight or when sparked. Explosive
$Cl_2(g)$	Substitution in diffused light → CH_3Cl, CH_2Cl_2, $CHCl_3$, CCl_4
Br_2(l or CCl_4)	CH_3Br in sunlight. Slow reaction
I_2	No reaction
$KMnO_4$, H_2SO_4	No reaction
$HNO_3(g)$	$C_2H_6 \rightarrow C_2H_5NO_2$ nitroethane (an industrial solvent)

Chlorination of methane

Initiation: Cl_2 + sunlight → $2Cl\cdot$

Propagating: $CH_3{-}H + Cl\cdot \rightarrow CH_3\cdot + HCl$, then $CH_3\cdot + Cl_2 \rightarrow CH_3Cl + Cl\cdot$

Terminating: $2Cl\cdot \rightarrow Cl_2$; $2CH_3\cdot \rightarrow C_2H_6$; $CH_3\cdot + Cl\cdot \rightarrow CH_3Cl$

Homologous series

Homologues have the same general formula and functional group. Molecular formulae differ by one CH_2, e.g. CH_4, CH_3CH_3, $CH_3CH_2CH_3$. They have similar preparations and chemical properties. Physical properties (m.p., b.p. and density) show gradual change up the series.

Isomers

The isomers of C_4H_{10} are $CH_3CH_2CH_2CH_3$ (butane) and $(CH_3)_2CHCH_3$ (methyl-propane).

Melting and boiling points

The m.p.s of methane and ethane are anomolous because their molecules have no zig-zag carbon chain and pack very closely. Larger molecules have greater temporary dipoles, greater attractive forces between them, and the m.p.s and b.p.s rise up the series.

QUESTIONS

1. State briefly the methods available for the preparation of methane. Name the reagents, state the conditions, and write an equation for each reaction you mention. Outline an experiment to distinguish between (a) methane and hydrogen, and (b) methane and butane.

2. By reference to the alkanes, explain what is meant by ***homologous series.*** How do the physical properties of alkanes vary with their position in the series?

3. The boiling points at 760 mmHg of three alkanes are: butane, 273 K; pentane, 309 K; hexane, 342 K. (a) Account for the fact that pentane has a higher boiling point than butane. (b) The boiling point of heptane is 341 K, 372 K or 399 K. Which figure do you think is correct? Explain your answer.

4. By reference to alkanes of molecular formula C_5H_{12}, explain the meaning of the term ***isomerism.*** Write the formulae of the isomers hexane and 2,2-dimethylbutane and state, with your reasons, which isomer has the higher boiling point.

5. The melting points at 760 mmHg of three alkanes are: pentane, 143 K; hexane, 178 K; heptane, 183 K. Predict the melting points of butane and octane. 2-methylpropane is an isomer of butane. Write its formula and state, with your reasons, which isomer has the lower melting point.

6. Explain what is meant by ***substitution***; refer only to the reaction between chlorine and ethane.

7. Describe briefly how petroleum is treated in order to obtain fuels and useful chemicals. In your answer refer to the fractionation of petroleum and to the cracking of substances obtained from petroleum.

8. Explain what is meant by ***free radical.*** Discuss the mechanism of the reaction between chlorine and methane.

Chapter 3
ALKENES AND RUBBER

ALKENES

The second homologous series of hydrocarbons is the alkenes. Their general formula is C_nH_{2n}, and $n = 2$ or more; the compound $CH_2 (n = 1)$ does not exist. The carbon atoms are in chains and not rings.

The names of alkenes end in *ene* and the first part of each name is the same as that for the corresponding alkane.

Name	*Formula*	*Relative molecular mass*	*B.p./K*	*M.p./K*
Ethene (ethylene)	C_2H_4	28	169.4	104.0
Propene (propylene)	C_3H_6	42	225.5	87.9
But-l-ene	C_4H_8	56	266.9	87.8
Pent-l-ene	C_5H_{10}	70	303	
Dec-l-ene	$C_{10}H_{20}$	140	445	

Two older names are in brackets. They are well established in commerce and everyday life.

Displayed (graphic) formulae and isomers of alkenes

All alkene molecules contain a double bond, consisting of four electrons or two electron pairs, between two carbon atoms:

$$\rangle C{=}C \langle$$

This *functional group* of alkenes determines their properties. The double bond accounts for the fact that an alkene molecule contains two hydrogen atoms less than an alkane molecule with the same number of carbon atoms, e.g. C_2H_4 and C_2H_6.

There are no isomers of the first two alkenes:

$CH_2{=}CH_2$ or $H_2C{=}CH_2$ or $CH_2{=}CH_2$ (written vertically) or $H{-}C{-}H = H{-}C{-}H$ or $H_2C{=}CH_2$ (with all bonds shown)

ethene (ethylene)

$CH_3{-}CH{=}CH_2$ or $H_3C{-}CH{=}CH_2$ or $CH_3{-}CH{=}CH_2$ (written vertically) or $H_3C(H)C{=}CH_2$ (with all bonds shown)

propene (propylene)

All six atoms of an ethene molecule are in one plane and the molecule is symmetrical.

Three isomers have the formula C_4H_8. Molecules of two have a chain of four carbon atoms, but the position of the double bond is different: the third isomer has a chain of three carbon atoms plus a branched chain:

C–C–C=C   C–C=C–C   C–C(–C)=C or C–C(=C)–C

The position of the double bond is located by a number. The numbering of the longest chain of carbon atoms starts from the end of the chain nearer to the double bond. The names and structural formulae are:

4 3 2 1 $CH_3CH_2CH{=}CH_2$ or $CH_2{=}CHCH_2CH_3$ 1 2 3 4	1 2 $CH_3CH{=}CHCH_3$	1 2 $CH_3C(CH_3){=}CH_2$
but-l-ene	but-2-ene	2-methylpropene

If there are two double bonds in a hydrocarbon, the letter '*a*' is added to the hydrocarbon root and the name ends in *-diene.* For example:

1 2 3 4
$CH_2{=}CH{-}CH{=}CH_2$ is buta-1,3-diene (not but-1,3-diene)

The numbers -1,3- mean that double bonds are between carbon atoms 1 and 2 and atoms 3 and 4. Make models of the molecules mentioned in this section.

Structure and shape of ethene molecules

A covalent bond consists of two electrons shared between atoms. There are two covalent bonds between the carbon atoms in ethene, C=C, but they are different. The electrons of one bond move in an orbital along the line joining the two carbon nuclei and it is called a σ-bond (sigma bond). (The four C—H bonds are also σ-bonds, and all five bonds are in one plane.) The two electrons of the other bond between the carbon atoms are not right between the atoms. The electron charge cloud lies above and below the line joining the two carbon nuclei. It is called a π-bond (pi bond), represented by the two sausage-shaped clouds in Fig. 3.1. The orbital is at right angles to the plane of the five σ-orbitals.

The electron charge clouds in a π-bond overlap less than in a σ-bond. The bonds are not of equal strength; the π-bond is weaker. The π-bond also prevents free rotation of the two carbon atoms about the bond. Atoms linked only by a σ-bond can rotate freely.

Ethane and other alkanes are reactive because the π-electrons are less firmly held by the nuclei than the σ-electrons. In an addition reaction of ethene, the π-bond is replaced by one new σ-bond on each carbon atom.

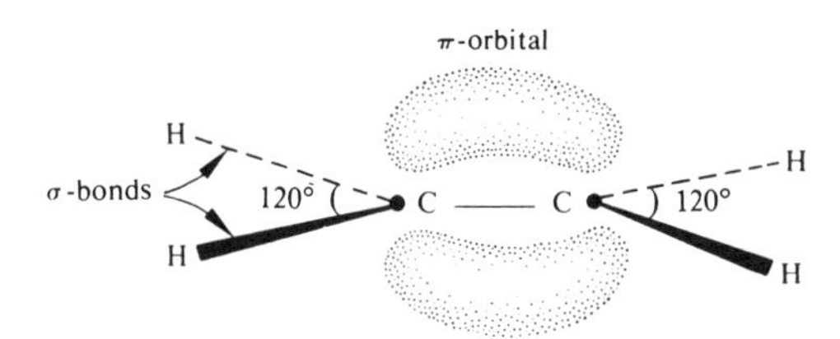

Fig. 3.1. Bonds in an ethene molecule. One σ-bond and one π-bond between the two carbon atoms

Laboratory preparation of ethene (ethylene)

Ethene is prepared by dehydration of (removal of the elements of water from) ethanol. Dehydration occurs when ethanol vapour is passed over hot broken porcelain or pumice stone or aluminium oxide, Al_2O_3, at 670 K, and also when ethanol is heated with either concentrated sulphuric acid at 440 K or with phosphoric(V) acid, H_3PO_4, at 470 K:

$$CH_3CH_2OH \rightarrow C_2H_4 + H_2O, \qquad \text{or} \qquad \begin{matrix} CH_3 \\ | \\ CH_2OH \end{matrix} \rightarrow \begin{matrix} CH_2 \\ \| \\ CH_2 \end{matrix} + \begin{matrix} H \\ | \\ OH \end{matrix}$$

1. *Dehydration of ethanol by porcelain or pumice stone.* Place a loose plug of glass wool or asbestos wool at the bottom of a boiling tube. Add ethanol slowly until the wool is soaked with the alcohol; pour out any ethanol not absorbed by

the wool. Fill the rest of the tube with pieces of broken porcelain or pumice stone. Fit the tube with a stopper and a delivery tube leading to a gas-jar inverted in a trough of water.

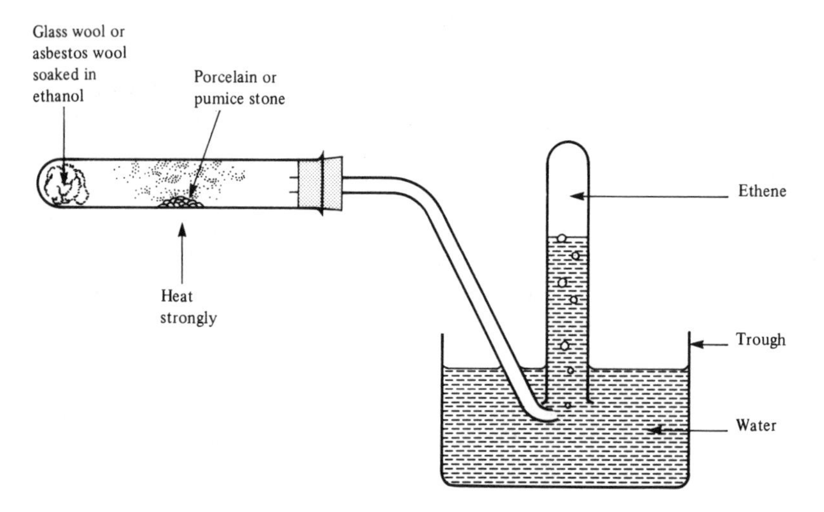

Fig. 3.2. Preparation of ethene by dehydration of ethanol

Heat the porcelain or pumice strongly. Some heat is conducted to the ethanol, which vaporizes steadily. The vapour passes through the heated porcelain or pumice and dehydration occurs. Collect the ethene over water.

2. *Dehydration of ethanol by aluminium oxide.* Activated aluminium oxide, prepared by heating hydrated aluminium hydroxide at a moderate temperature, is the catalyst. It is very porous and has a high surface area (up to 300 square metres per gram). The surface of the oxide has a great affinity for water, and makes it an efficient dehydration catalyst.

Pack hard-glass tubing (about 120 mm long and 20 mm diameter) with small granules of aluminium oxide and hold them in place with glass wool. Add about 10 cm^3 of ethanol to a flask. Arrange the apparatus as in Fig. 3.3. (Fig. 3.2 apparatus is simpler for preparing a little ethene.)

Heat the oxide with the blue flame of a bunsen burner to about 670 K (well below red heat). Heat the ethanol until its vapour passes in a steady stream through the oxide. Collect the ethene over water. If the temperature is below 530 K the main product is ethoxyethane (diethyl ether):

$$C_2H_5OH \xrightarrow[670\,K]{Al_2O_3} CH_2{=}CH_2; \qquad 2C_2H_5OH \xrightarrow[500\,K]{Al_2O_3} C_2H_5OC_2H_5$$

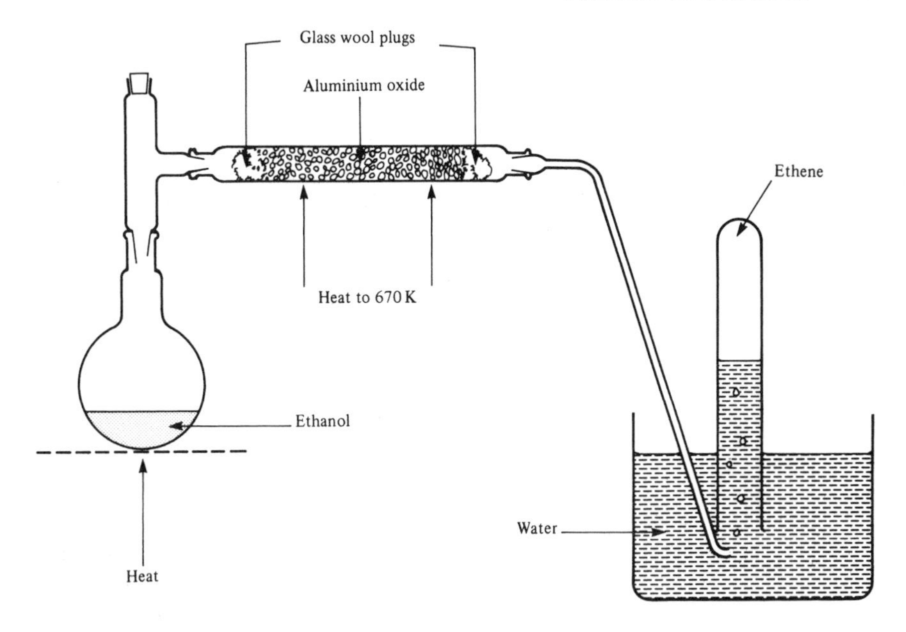

Fig. 3.3. Preparation of ethene from ethanol, using aluminium oxide as catalyst

3. *Dehydration of ethanol by excess concentrated sulphuric acid.* Add 100 cm^3 of concentrated sulphuric acid carefully and in small volumes at a time to 50 cm^3 of ethanol in a $1\frac{1}{2}$- or 2-litre flask. Much heat is evolved. Shake during the addition and cool under a tap or in a trough of water to prevent loss of ethanol by evaporation. Each ethanol molecule adds one proton, H^+, from the acid:

$$C_2H_5OH(l) + H_2SO_4(l) \rightleftharpoons HSO_4^- + \underset{\text{protonated ethanol}}{C_2H_5OH_2^+}$$

Add a little sand to the mixture to prevent or reduce frothing during subsequent heating. Arrange the apparatus as in Fig. 3.4.

Heat the flask on a sand bath, which soaks up the corrosive acid mixture if the flask accidentally breaks. Ethene is evolved at about 440 K:

$$C_2H_5OH_2^+ \rightleftharpoons C_2H_4(g) + H^+ + H_2O$$

$$\left(C_2H_5OH \xrightarrow[440\text{ K}]{\text{excess } H_2SO_4} C_2H_4 \right)$$

Both stages are reversible.

Some frothing occurs but all the froth should remain in the large flask. The liquid turns black owing to formation of carbon, which is usually produced when organic compounds are heated with sulphuric acid:

$$C_2H_5OH(l) + 2H_2SO_4(l) \rightarrow 2C(s) + 2SO_2(g) + 5H_2O$$

The gas evolved contains carbon dioxide and sulphur dioxide, formed by reduction of the acid by carbon:

$$C(s) + 2H_2SO_4(l) \rightarrow CO_2(g) + 2SO_2(g) + 2H_2O$$

Remove the two dioxides by passing the gas through a wash-bottle containing 10 per cent aqueous sodium hydroxide. Collect 9 tubes or jars of the gas. Ethanol vapour and ethoxyethane present in the gas evolved are removed by the water or the alkali solution.

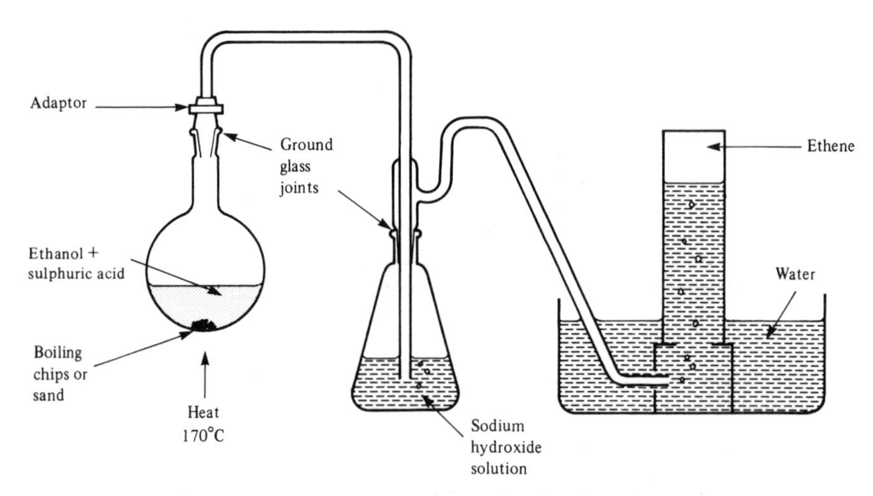

Fig. 3.4. Preparation of ethene from ethanol, using sulphuric acid

If a large volume of ethene is required, the flask should be fitted with a dropping funnel; equal volumes of ethanol and concentrated sulphuric acid should be mixed carefully in a beaker and added to the flask from the funnel at a rate sufficient to maintain the flow of gas.

4. *Dehydration of ethanol by phosphoric(V) acid,* H_3PO_4. Pure ethene is obtained by using phosphoric acid instead of sulphuric acid. Add 100 cm^3 of phosphoric acid to a flask and 40 cm^3 of ethanol to a dropping funnel in the neck of the flask. Heat the acid to about 470 K on a sand bath and add the ethanol drop by drop. Collect the ethene over water. The apparatus is similar to that of Fig. 3.4 without the wash bottle.

5. *Elimination of hydrogen halide from halogenoalkanes.* Propene is prepared by heating 2-bromopropane with alcoholic potassium hydroxide (potassium

hydroxide dissolved in methylated spirit, ethanol or pentanol). The alkali removes or 'eliminates' the elements of hydrogen bromide from the halogen compound. (A reaction which effects the removal of a simple compound and the formation of a double bond or triple bond in a compound is called an *elimination reaction.*)

$$\begin{array}{c} CH_3CHBr \\ | \\ CH_3 \end{array} \xrightarrow{KOH} \begin{array}{c} Br \\ | \\ H \end{array} + \begin{array}{c} CH_3CH \\ \| \\ CH_2 \end{array}$$

propene

or $CH_3CHBrCH_3(l) + KOH(alc) \rightarrow KBr + H_2O + CH_3CH{=}CH_2(g)$

Ethene cannot be prepared by this method because bromoethane and alcoholic potassium hydroxide react to form mainly ethoxyethane (diethyl ether) and the yield of ethene is only about 2 per cent:

$$C_2H_5Br + KOH(alc) \rightarrow KBr + H_2O + C_2H_4 \text{ (2 per cent only)}$$

$$C_2H_5Br + HOC_2H_5 \rightarrow HBr + C_2H_5OC_2H_5$$

Tests on ethene

1–3. Do the combustion, bromine and potassium manganate(VII) tests as for methane (p. 20).

4. *Chlorine.* Place a gas-jar of chlorine mouth to mouth over a jar of ethene. Allow the two gases to mix. Note if the greenish-yellow colour of chlorine remains.

Properties of ethene

Physical properties. Ethene is a colourless gas with a sweetish smell. It is sparingly soluble in water and more soluble in ethanol. The boiling points and melting points of alkenes rise with increasing relative molecular mass because of greater van der Waals' forces between the larger molecules which are more easily polarized.

Chemical properties

1. *Combustion.* Ethene burns in air or oxygen with a luminous smoky flame. The flame is smoky because of the high carbon content (85.7 per cent by mass).

$$C_2H_4 + 3O_2 \rightarrow 2CO_2 + 2H_2O, \quad \text{and also}$$

$$C_2H_4 + 2O_2 \rightarrow 2CO + 2H_2O$$

$$C_2H_4 + O_2 \rightarrow 2C + 2H_2O$$

Ethene burns in chlorine with a reddish flame, and clouds of soot form:

$$C_2H_4(g) + 2Cl_2(g) \rightarrow 4HCl(g) + 2C(s)$$

2. *Addition reactions of ethene.* Two carbon atoms in an alkene molecule do not exert their maximum combining powers and there is a double bond, C=C, between them. Alkenes are therefore *unsaturated* compounds. A molecule XY can 'add' to an alkene molecule to form one molecule of a saturated compound. *Addition* takes place across the double bond; the product is an *addition compound*, and the reaction is an *addition reaction*. The general equation is:

$$\begin{array}{c}CH_2\\ \|\\ CH_2\end{array} + \begin{array}{c}X\\ |\\ Y\end{array} \rightarrow \begin{array}{c}CH_2X\\ |\\ CH_2Y\end{array} \quad \text{or} \quad \begin{array}{c}H\diagdown\ \diagup H\\ C\\ \|\\ C\\ H\diagup\ \diagdown H\end{array} + \begin{array}{c}X\\ |\\ Y\end{array} \rightarrow \begin{array}{c}H\\ |\\ H-C-X\\ |\\ H-C-Y\\ |\\ H\end{array}$$

X and Y are univalent atoms or groups and they may be the same.

(a) *Hydrogen.* Hydrogen gas and ethene combine in the presence of a catalyst. Finely divided platinum or palladium catalyze the addition (hydrogenation) at ordinary temperatures and pressure; nickel catalyses the reaction under pressure at about 470 K:

$$\begin{array}{c}H\diagdown\ \ \diagup H\\ C=C\\ H\diagup\ \ \diagdown H\end{array} + H_2 \xrightarrow[\text{(Ni, Pt, Pd)}]{\text{catalyst}} \begin{array}{c}H\ \ \ H\\ |\ \ \ \ |\\ H-C-C-H\\ |\ \ \ \ |\\ H\ \ \ H\end{array}$$

(b) *Halogens.* Fluorine reacts violently with ethene. Chlorine gas, liquid bromine, a solution of bromine in tetrachloromethane, and a solution of iodine in ethanol all add readily to ethene. The products are colourless oils. No hydrogen halide is formed. Bromine forms 1,2-dibromoethane.

$$\begin{array}{c}CH_2\\ \|\\ CH_2\end{array} + \begin{array}{c}Br\\ |\\ Br\end{array} \rightarrow \begin{array}{c}CH_2Br\\ |\\ CH_2Br\end{array} ; \quad \begin{array}{c}H\diagdown\ \ \diagup H\\ C=C\\ H\diagup\ \ \diagdown H\end{array} + \begin{array}{c}Br\\ |\\ Br\end{array} \longrightarrow \begin{array}{c}H\ \ \ H\\ |\ \ \ \ |\\ H-C-C-H\\ |\ \ \ \ |\\ Br\ \ Br\end{array}$$

The order of reactivity with alkenes of the halogens is F > Cl > Br > I. Addition of chlorine or bromine to an unsaturated compound is called *chlorination* or *bromination* (*halogenation*).

(c) *Hydrogen halides.* Ethene reacts readily at ordinary temperatures with hydrogen iodide and bromide and with concentrated hydriodic and hydrobromic acids, but it reacts very slowly with hydrochloric acid. Ethene and hydrogen chloride react at 420 K with aluminium chloride as a catalyst.

$$CH_2{=}CH_2 + H{-}Br(\text{g or aq}) \rightarrow CH_3CH_2Br; \qquad \begin{matrix} CH_2 \\ \| \\ CH_2 \end{matrix} + \begin{matrix} H \\ | \\ I \end{matrix} \rightarrow \begin{matrix} CH_3 \\ | \\ CH_2I \end{matrix}$$

Two reactions are possible with propene and hydrogen halides:

$$CH_3CH{=}CH_2 + H{-}Br \rightarrow \underset{\text{1-bromopropane}}{CH_3CH_2CH_2Br} \quad \text{or} \quad \underset{\text{2-bromopropane}}{CH_3CHBrCH_3}$$

The product is the 2-halogeno-compound. In these and similar addition reactions the hydrogen atom joins to the alkene carbon atom which already has the larger number of hydrogen atoms. This rule applies when a hydrogen compound adds to an alkene or alkyne with an unsymmetrical molecule. It is called *Markownikoff's rule.*

$$RCH{=}CH_2 + HBr \rightarrow \begin{matrix} RCHCH_3 \\ | \\ Br \end{matrix}$$

The order of reactivity of the hydrogen halides is $HI > HBr > HCl > HF$; contrast this order with that of the halogens.

(d) *Chloric(I) acid and bromic(I) acid,* $HClO$ *and* $HBrO$. These acids are in chlorine water and bromine water respectively. Two products are formed when ethene reacts with each halogen water:

$$\begin{matrix} CH_2 \\ \| \\ CH_2 \end{matrix} + \begin{matrix} Cl \\ | \\ Cl \end{matrix} \rightarrow \begin{matrix} CH_2Cl \\ | \\ CH_2Cl \end{matrix}; \qquad \begin{matrix} CH_2 \\ \| \\ CH_2 \end{matrix} + \begin{matrix} OH \\ | \\ Cl \end{matrix} \rightarrow \begin{matrix} CH_2OH \\ | \\ CH_2Cl \end{matrix}$$

The products are 1,2-dichloroethane and 2-chloroethanol. Bromine water forms similar bromo-products.

(e) *Sulphuric acid.* At room temperature the fuming acid adds rapidly to ethene and the concentrated acid adds slowly forming ethyl hydrogensulphate:

$$C_2H_4 + H_2SO_4 \rightarrow C_2H_5HSO_4 \quad \text{or} \quad \begin{matrix} CH_2 \\ \| \\ CH_2 \end{matrix} + \begin{matrix} H \\ | \\ OSO_2OH \end{matrix} \rightarrow \begin{matrix} CH_3 \\ | \\ CH_2OSO_2OH \end{matrix}$$

This reaction is used in industry to separate alkenes from alkanes in the products obtained when petroleum is 'cracked'.

(f) *Water*. Ethene and steam combine at 600 K and about 70 atm pressure in the presence of phosphoric(V) acid as catalyst:

$$C_2H_4 + H_2O \rightarrow C_2H_5OH \quad \text{or} \quad \begin{matrix} CH_2 \\ \| \\ CH_2 \end{matrix} + \begin{matrix} H \\ | \\ OH \end{matrix} \rightarrow \begin{matrix} CH_3 \\ | \\ CH_2OH \end{matrix}$$

This reaction is used in the modern process for producing industrial ethanol, which is also manufactured by absorbing ethene in concentrated sulphuric acid at 373 K and adding water to the product. The reactions are the reverse of those on p. 41.

The addition of the elements of water across a double or triple bond is a *hydration reaction.*

(g) *Oxygen*. Ethene and oxygen or air combine under pressure at 500 K, with a silver catalyst, to form epoxyethane. This is hydrolysed by hot water, and the reaction is used to manufacture ethane-1,2-diol:

$$\begin{matrix} CH_2 \\ \| \\ CH_2 \end{matrix} \xrightarrow{O_2} \begin{matrix} H_2C \\ | \\ H_2C \end{matrix}\!\!>\!O \xrightarrow{H_2O} \begin{matrix} CH_2OH \\ | \\ CH_2OH \end{matrix}$$

Ethanal is manufactured by the reaction between ethene and oxygen or air, using a special catalyst:

$$2CH_2{=}CH_2 + O_2 \rightarrow \underset{\text{ethanal}}{2CH_3CHO}$$

The mechanism of this oxidation is complicated.

(h) *Potassium manganate*(*VII*), $KMnO_4$. This reagent oxidizes ethene to a diol, which is a compound containing two hydroxyl groups.

$$\begin{matrix} CH_2 \\ \| \\ CH_2 \end{matrix} + [O] \text{ (from } KMnO_4) + \begin{matrix} H \\ | \\ OH \end{matrix} \rightarrow \underset{\text{ethane-1,2-diol}}{\begin{matrix} CH_2OH \\ | \\ CH_2OH \end{matrix}}$$

A pink acidified potassium manganate solution is decolorized; an alkaline manganate solution is reduced to green potassium manganate(VI), K_2MnO_4, and then brown hydrated manganese(IV) oxide is precipitated.

(i) *Hydrogen peroxide.* This adds to ethene when ethanoic acid is used as a solvent:

$$\begin{matrix} CH_2 \\ \| \\ CH_2 \end{matrix} + \begin{matrix} OH \\ | \\ OH \end{matrix} \rightarrow \begin{matrix} CH_2OH \\ | \\ CH_2OH \end{matrix}$$

(j) *Ozone*. Alkenes and ozone in an inert solvent (e.g. hexane, tetrachloromethane) form ozonides at ordinary temperatures. The double bond breaks and a ring compound forms:

$$CH_2{=}CH_2 + O_3 \rightarrow \text{ethene ozonide (ring: } O\langle CH_2{-}O,\ CH_2{-}O \rangle)$$

ethene ozonide

Addition of bromine to alkenes (electrophilic reactions)

From the equation for the reaction between ethene and bromine, it seems that the double bond of one ethene molecule breaks and one bromine molecule adds on to the free valency bonds so formed. This is not the true mechanism. The addition reaction takes place in three stages.

1. A bromine molecule approaches the ethene double bond. The electrons in the readily accessible π-orbital polarize the bromine molecule by causing a movement of an electron pair in the molecule. This movement is represented by a curved arrow; the arrow starts where the electron pair is before reaction and points to its destination:

$$Br{-}Br \quad (\text{i.e. } \overset{\delta+}{Br}{-}\overset{\delta-}{Br})$$

2. The ethene orbital electrons then attract the positive end of the polarized bromine molecule, which acts as an electrophile. A positive carbocation and a negative bromide ion are formed, and heterolysis of the bromine has occurred:

$$H_2C{=}CH_2 + Br{-}Br \longrightarrow H_2C{-}CH_2\ (\text{bridged by } Br^+) + Br^-$$

A *carbocation* is a positively charged species with the positive charge on carbon. A *carbanion* has a negative charge on the carbon.

3. A negative bromide ion (not necessarily the one just formed) now attacks the carbocation. The bromine already in the carbocation is large enough to keep away the bromide ion, which therefore attacks on the other side of the carbon-carbon bond. The negative charge of the bromide ion and the positive charge of the carbocation cancel each other.

$$Br^- + H_2C{-}CH_2\ (\text{bridged by } Br^+) \longrightarrow H{-}\underset{H}{\overset{Br}{C}}{-}\underset{Br}{\overset{H}{C}}{-}H$$

Since the bromine particles: Br^- and Br^+, enter the ethene molecule on opposite sides, the reaction is a *trans* addition (trans means 'on opposite sides of').

The characteristic reaction of alkenes (and alkynes) is electrophilic addition. The general equation, using E for an electrophile and Nu for a nucleophile, is:

$$\rangle C{=}C\langle + E{-}Nu \longrightarrow \rangle \overset{+}{C}{-}\underset{|}{\overset{E}{\overset{|}{C}}}{-} + :Nu^- \longrightarrow -\underset{Nu}{\underset{|}{\overset{|}{C}}}-\underset{|}{\overset{E}{\overset{|}{C}}}-$$

Bromine water and ethene. Polarization of bromine occurs as in stage 1 above, and a carbocation forms as in stage 2.

In stage 3, the carbocation can react with Br^- or H_2O, and two products are possible. If it reacts with Br^-, stage 3 is the same as above. If it reacts with H_2O, one lone pair of the oxygen atom forms a carbon–oxygen bond, and the product loses a proton (to another H_2O molecule) and yields 2-bromoethanol:

$$H_2\ddot{O}: + {}^{+}CH_2{-}CH_2Br \longrightarrow H_2\overset{+}{O}{-}CH_2{-}CH_2Br \xrightarrow{\text{proton loss to } H_2O} HO{-}CH_2{-}CH_2{-}Br + H_3O^+$$

The fact that reaction between ethene and bromine water yields 2-bromoethanol in addition to the expected 1,2-dibromoethane proves that reaction does not take place in one step by addition of one bromine molecule to one ethene molecule. Similarly, bromine water containing some sodium chloride also forms a third product, 1-bromo-2-chloroethane, $ClCH_2CH_2Br$. This supports the formation of a carbocation $CH_2CH_2Br^+$ as a first step, followed by addition of any one of the three nucleophiles present in the reaction mixture: OH^-, Br^- or Cl^-.

Mechanism of addition of hydrogen bromide to ethene

A molecule of hydrogen bromide is polarized: $\overset{\delta+}{H}{-}\overset{\delta-}{Br}$. The positively charged hydrogen atom is an electrophile and is attracted to the π-electrons of the double bond in ethene:

$$H_2C{=}CH_2 + \overset{+}{H}{-}\overset{-}{Br} \rightarrow CH_3\overset{+}{C}H_2 + Br^- \quad \text{(slow)}$$

This first stage is similar to the second stage with Br_2 on p. 49.

The carbocation then reacts with a bromide ion (a nucleophile):

$$CH_3\overset{+}{C}H_2 + Br^- \rightarrow CH_3CH_2Br \quad \text{(rapid)}$$

Tests for unsaturation

The reactions with bromine and acidified potassium manganate(VII) are tests for unsaturation in organic compounds. A reddish-brown solution of bromine in tetrachloromethane and a pink aqueous solution of potassium manganate(VII), acidified with dilute sulphuric acid, are decolorized rapidly by alkenes and other unsaturated compounds.

Catalytic hydrogenation of ethene

Finely divided nickel, which has a large active surface, is the catalyst used commercially when compounds with an alkene double bond are hydrogenated. The atoms inside the nickel are surrounded by and joined to other atoms. The surface atoms have free valency bonds. Both ethene and hydrogen are adsorbed by nickel. Since new bonds are formed between the surface nickel atoms and the two gases, the process is called *chemisorption.* The chemisorbed gases react:

$$C_2H_4\,(\text{chemisorbed}) + H_2\,(\text{chemisorbed}) \rightarrow C_2H_6(g)$$

When hydrogen is chemisorbed by nickel, the hydrogen molecules dissociate into atoms, each of which forms a covalent bond with a surface nickel atom. An ethene molecule is chemisorbed by the opening of its double bond and the formation of two covalent bonds with surface nickel atoms:

$$-\text{Ni}-\text{Ni}- + H_2 \rightarrow -\underset{}{\overset{\displaystyle H\ \ \ \ H}{\overset{|\ \ \ \ \ |}{\text{Ni}-\text{Ni}}}}-$$

$$-\text{Ni}-\text{Ni}- + H_2C{=}CH_2 \rightarrow -\overset{\displaystyle H_2C-CH_2}{\overset{|\ \ \ \ \ |}{\text{Ni}-\text{Ni}}}-$$

Experimental evidence indicates that hydrogen atoms add one at a time to the chemisorbed ethene, forming first a chemisorbed ethyl radical, thus:

$$C_2H_4\,(\text{chemisorbed}) \xrightarrow{+H\cdot} C_2H_5\cdot(\text{chemisorbed}) \xrightarrow{+H\cdot} C_2H_6(g)$$

Nickel, platinum and palladium are good hydrogenation catalysts because they readily chemisorb gases. Atoms of these metals have single (unpaired) electrons, and these electrons are necessary for chemisorption.

Uses of alkenes

Alkenes are produced commercially by cracking either petroleum or some fractions from petroleum. Therefore all substances obtained from alkenes are derived from petroleum.

Ethene is used to ripen some fruits while they are transported in ships; the gas stimulates production of enzymes responsible for ripening.

Ethanol. Direct hydration of ethene with steam is used (p. 48) and also indirect hydration using sulphuric acid and then water (p. 48).

Epoxyethane. Ethene and oxygen combine directly (p. 48). The product, $(CH_2)_2O$, is used as a fumigant for foodstuffs and tobacco, and to manufacture ethane-1,2-diol.

Ethane-1,2-diol. Epoxyethane is heated with water under pressure (p. 48).

Other products obtained from ethene include halogenoalkanes (p. 99) poly(ethene) (p. 53), poly(phenylethene) (p. 82) and ethanal (p. 48).

Propan-2-ol. Propene is hydrated directly at high temperature and pressure in the presence of a catalyst:

$$CH_3CH{=}CH_2 + H_2O \rightarrow CH_3CHOHCH_3$$

Propene is also hydrated indirectly by liquefying the gas under pressure and then absorbing the liquid in concentrated sulphuric acid. The reaction mixture is diluted with water and heated.

Detergents. These are manufactured from alkenes whose molecules contain from 12 to 20 carbon atoms (p. 216).

Polymerization of ethene

Ethene molecules can combine under certain conditions and form white solid *polyethylene,* sometimes known as *polythene.* The systematic name is poly(ethene). The reaction depends on the ability of carbon atoms to form long chains. As many as 700 ethene molecules combine to form one large molecule, a *macromolecule*, containing a chain of about 1400 carbon atoms. Ethene is a *monomer* (one part) and poly(ethene) is a *polymer* (many parts) or a *high polymer*. Ethene molecules combine because one molecule can add to another owing to the double bond present. *Addition polymerization* is the combination of single molecules to form one molecule without loss or gain of atoms. The monomer and the polymer therefore have the same empirical formula. The formula of poly(ethene) is $(C_2H_4)_n$ where n is a large number up to about 350.

$$nCH_2{=}CH_2 \rightarrow \underset{\text{poly(ethene)}}{+CH_2{-}CH_2+}$$

There is a CH_3 group at each end of the carbon chain.

The carbon chain is zig-zag because of the tetrahedral distribution of the bonds of each carbon atom (p.9). A better formula for poly(ethene) is:

```
        H   H   H   H                        H H    H H
        |   |   |   |            or           \ /    \ /
 --- —C—C—C—C— ----                   \   C    \   C   \
        |   |   |   |                       C      C
        H   H   H   H                      / \    / \
                                          H   H  H   H
```

The carbon atoms are in the plane of the paper. Each thick line represents a covalent bond coming out of the paper and a broken line represents a bond going down through the paper away from you. A three-dimensional model makes the structure clear.

Poly(ethene) (polyethylene or polythene)

Manufacture

1. *High pressure process.* Ethene is heated to 470 K under 1500 atm pressure with a trace of oxygen as catalyst. The product is called low density poly(ethene). It contains some branched chain molecules, which cannot pack closely in the solid and therefore account for the relatively low density of 0.92 g cm^{-3}.

2. *Ziegler process.* Ethene is dissolved in a hydrocarbon solvent containing in suspension a catalyst of triethylaluminium, $Al(C_2H_5)_3$ and titanium(IV) chloride, $TiCl_4$. The temperature is about 340 K and the pressure 2 to 7 atm. Water or dilute acid is added after reaction to decompose the catalyst, and the poly(ethene) is recovered from the mixture by filtration. The product is called high density poly(ethene); its density is 0.95 g cm^{-3}. It has a higher tensile strength and softening temperature (about 405 K) than low density poly(ethene).

Processes similar to the Ziegler process use different catalysts, temperatures and pressures.

The products made by the two processes differ. That made at high pressure has many side chains, e.g.

```
—C—C—C—C—          etc.
        |
        C
        |
        C
```

These side chains prevent close packing of the molecules and this poly(ethene) has a low density and a low softening point, about 95 °C. It is called LDPE which means low-density poly(ethene). The product made at lower pressures and with special catalysts has fewer side chains and the mainly linear molecules can pack together closely. It has a higher density and softening point, about 120 °C, and is called HDPE, high-density poly(ethene).

Properties. Poly(ethene) is a white waxy solid. It is less dense than water, tough, insoluble in all common solvents, not affected by acids or alkalis or common chemical reagents, does not rot or corrode, and is a good electrical insulator. It softens on warming and does not have a definite melting point because its molecules vary in length and mass and slightly in chemical composition.

Uses. The sheet plastic is used to make carrier bags, containers for cans, bottles and cartons, and sacks to hold various substances such as fertilizers, foodstuffs, household rubbish, etc. It is used to cover growing plants to protect them from bad weather. Ropes, carpets, water pipes, toys, dustbins, etc. are made from HDPE. It is used as an insulator around electrical cables, and especially for telephone cables under the ground or sea.

Polymers from substituted ethene compounds

The four atoms joined to carbon in an ethene molecule are all hydrogen. Chemists have changed one or more of these atoms, polymerized the product, and obtained polymers with the kind of properties required in everyday life. Chloroethene (vinyl chloride), $CH_2{=}CHCl$, and phenylethene (styrene), $CH_2{=}CHC_6H_5$, polymerize to form poly(chloroethene), commonly called polyvinyl chloride or PVC, and poly(phenylethene) or polystyrene. The formulae are:

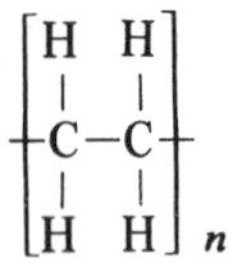

$$\left[\begin{array}{c} H\quad H \\ |\quad\ | \\ -C-C- \\ |\quad\ | \\ H\quad H \end{array}\right]_n \qquad \left[\begin{array}{c} H\quad Cl \\ |\quad\ | \\ -C-C- \\ |\quad\ | \\ H\quad H \end{array}\right]_n \qquad \left[\begin{array}{c} H\quad C_6H_5 \\ |\quad\ | \\ -C-C- \\ |\quad\ | \\ H\quad H \end{array}\right]_n$$

poly(ethene) polythene — poly(chloroethene) PVC — poly(phenylethene) polystyrene

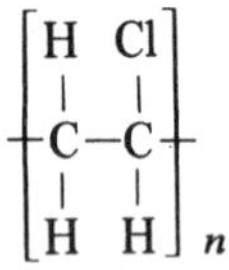

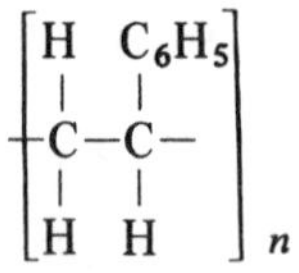

Poly(chloroethene) or polyvinyl chloride, PVC

Chloroethene, $CH_2{=}CHCl$, polymerizes more readily than ethene when heated with a suitable catalyst. Addition of suitable chemicals can produce a hard rigid PVC or a soft flexible product. The hard kind is used for gutters on roofs, light fittings, and water pipes. Soft PVC is used for an insulating covering over electrical wires, raincoats, waterproof shoes, and 'artificial leather' in cars.

$$nCH_2{=}CHCl \rightarrow {+}CH_2{-}CHCl{+}_n$$

Poly(phenylethene) or polystyrene

Phenylethene, $CH_2{=}CHC_6H_5$, also polymerizes readily when heated with a suitable catalyst:

$$n CH_2{=}CHC_6H_5 \rightarrow \left[CH_2{-}CHC_6H_5 \right]_n \quad n = \text{about } 5000$$

The large phenyl groups, C_6H_5, prevent the molecules from vibrating and bending easily, and account for the relatively high softening point, 230 °C, of the plastic.

It does not rot or corrode, resists chemical reagents, and is a good insulator. It is used to make containers and tableware and for electrical and thermal insulation. In the form of a foam it is widely used for packaging fragile goods.

The arrangements of atoms in the macromolecules of poly(chloroethene) and poly(phenylethene) are:

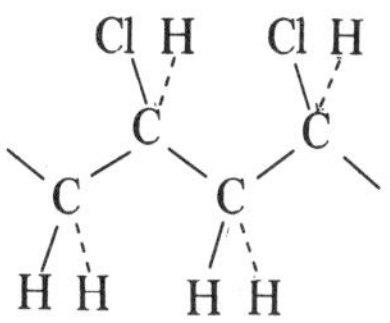

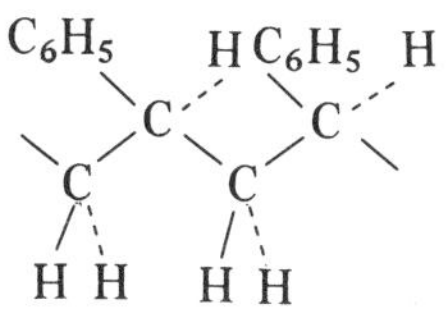

```
        Cl H    Cl H                  C6H5     H C6H5  H
         \/      \/                       \    C      \  C
    \    C   /\   C   \               \  /  \  /  \  /  \
     C  /    C  /                      C      C
    / \      / \                      / \    / \
   H   H    H   H                    H   H  H   H
```

To polymerize phenylethene. To a test-tube in a fume cupboard add 5 cm^3 of phenylethene, 5 cm^3 of 1,2-dibromoethane, and about 0.1 g of lauroyl peroxide, a catalyst. Place a wad of glass wool loosely in the mouth of the tube, and leave the tube in a beaker of hot water for about 20 minutes. Allow to cool, add about 10 cm^3 of propanone, and decant the solution from the solid residue that forms. To the solution add excess ethanol, which precipitates poly(phenylethene) as a white solid.

RUBBER

Natural rubber is a polymer with long carbon chains:

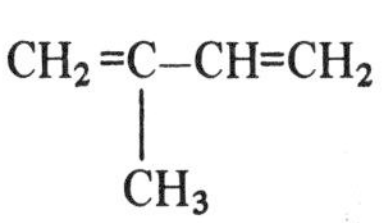

$$CH_2{=}\underset{\displaystyle CH_3}{\underset{|}{C}}{-}CH{=}CH_2$$

2-methyl-1,3-diene

$$\left[CH_2{-}\underset{\displaystyle CH_3}{\underset{|}{C}}{=}CH{-}CH_2 \right]_n$$

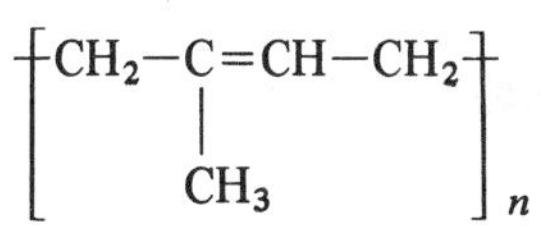

natural rubber (n = 1000 to 5000)

Hot rubber reacts with sulphur to form strong, elastic commercial rubber. The sulphur adds to a few double bonds (about 1 in 75). The carbon chains are now cross-linked by carbon–carbon and carbon–sulphur covalent bonds:

$$\begin{array}{l} \text{+}CH_2\text{—}C(CH_3)\text{=}CHCH\text{+}_m \\ \qquad\qquad\qquad\qquad\quad | \\ \quad \text{+}CH_2\text{—}C(CH_3)\text{—}CHCH_2\text{+}_n \\ \qquad\qquad\;\; | \\ \qquad\qquad\;\; S \\ \qquad\qquad\;\; | \end{array}$$

commercial rubber

Molecules of poly(ethene), poly(phenylethene), and other polyalkenes, which have no cross-links, are held in place only by relatively weak van der Waals' forces (p. 27). At high temperatures these forces are not strong enough; the polymers soften on heating but solidify on cooling. The change is physical and reversible. A plastic which can be softened without decomposition is *thermoplastic.* Commercial rubber (and Bakelite) are called *thermosetting* plastics because each softens and decomposes on heating. Heat breaks the covalent cross-links in the rubber molecules. The change is chemical and irreversible.

SUMMARY

Alkenes

C_nH_{2n}. Functional group: C=C. Electrophilic reagents attack the weak π-bond.

Preparation

1. Dehydrate ethanol with hot porcelain or pumice stone, Al_2O_3, H_2SO_4(conc) at 440 K, or H_3PO_4 at 470 K.

2. Halogenoalkane and KOH(alc)

$$CH_3CHBrCH_3 \rightarrow \underset{\text{propene}}{CH_3CH{=}CH_2}$$

Properties
(of C_2H_4 or $CH_2{=}CH_2$, ethene)

Combustion:	
Air	$CO_2 + H_2O$. Sooty, luminous flame
Chlorine	$C + HCl$. Sooty, reddish flame
Addition:	
H_2/Ni or Pt	C_2H_6 or CH_3CH_3 ethane
$Cl_2(g)$	$CH_2ClCH_2Cl(l)$ 1,2-dichloroethane
Br_2(l or CCl_4)	$CH_2BrCH_2Br(l)$ Cl > Br > I 1,2-dibromoethane
HBr(g or aq)	CH_3CH_2Br HI > HBr > HCl bromoethane
$HCl/AlCl_3$, 150°C	CH_3CH_2Cl chloroethane
Steam/H_3PO_4	CH_3CH_2OH Industrial process ethanol
O_2/Ag, pressure	C_2H_4O epoxyethane (a ring compound) Water forms CH_2OHCH_2OH Industrial process ethane-1,2-diol
$KMnO_4/H^+(aq)$	Decolorized CH_2OHCH_2OH
H_2O_2	CH_2OHCH_2OH
O_3(hexane)	$C_2H_4O_3$ ethene ozonide (a ring compound)

Mechanism of addition
Addition of Br_2:

1. Br_2 polarized by π-electrons, $Br_2 \rightarrow \overset{+}{Br}{-}\overset{-}{Br}$
2. $H_2C{=}CH_2 + \overset{+}{Br}{-}\overset{-}{Br} \rightarrow H_2C\overset{+}{Br}CH_2 + Br^-$
3. $H_2C\overset{+}{Br}CH_2 + Br^- \rightarrow CH_2BrCH_2Br$ (Br^+ and Br^- enter opposite sides of molecule)

Addition of HBr:

1. $CH_2{=}CH_2 + \overset{+}{H}{-}\overset{-}{Br}$ (polarized) $\rightarrow CH_3\overset{+}{C}H_2 + Br^-$

2. $CH_3\overset{+}{C}H_2 + Br^- \rightarrow CH_3CH_2Br$

Polymers

Monomer:	$CH_2{=}CH_2$ ethene	$CH_2{=}CHCl$ chloroethene	$CH_2{=}CHC_6H_5$ phenylethene
Polymer:	poly(ethene)	poly(chloroethene)	PVC poly(phenylethene)
Repeat unit:	$(-CH_2-CH_2-)$	$(-CH_2-CHCl-)$	$(-CH_2-CHC_6H_5-)$

The above plastics are thermoplastic. Rubber and Bakelite are thermosetting.

QUESTIONS

1. By naming the reagents, stating the conditions, and writing an equation for each reaction involved, outline four methods of preparing ethene in the laboratory.

2. Write the displayed formula of ethene and of two C_4H_8 alkenes.

3. Name and write the displayed formulae of isomeric alkenes C_4H_8. Predict, with reasons, which one has the highest boiling point.

4. Ethene is an ***unsaturated*** compound. By referring to its reactions with liquid bromine, bromine water and hydrogen bromide, explain the meaning of this statement.

5. What is meant by ***addition polymerization***? Describe briefly an experiment by which you have prepared one polymer of an organic compound in the laboratory.

6. Discuss the mechanism of the addition reaction that occurs when liquid bromine and ethene combine.

7. Draw structural formulae and give systematic names of any four alkenes of molecular formula C_6H_{12}. What reactions are possible between but-1-ene, $CH_3CH_2CH{=}CH_2$, and (a) hydrogen, (b) hydrogen bromide, and (c) concentrated sulphuric acid followed by water?

Chapter 4
ALKYNES

The alkynes, formerly called the *acetylenes,* is a third homologous series of hydrocarbons. Their general formula is C_nH_{2n-2}, and $n = 2$ or more. The carbon atoms are in chains.

The names of alkynes end in *yne* and the first part of each name is the same as that for the corresponding alkane. The first member of the series is ethyne (acetylene). The higher homologues (propyne, butyne, etc.) are of little commercial importance and therefore this chapter deals with ethyne.

Displayed (graphic) formulae of alkynes

All alkyne molecules contain a triple bond, consisting of six electrons or three electron pairs, between two carbon atoms: $-C{\equiv}C-$. This *functional group* of alkynes determines their properties. The triple bond accounts for the fact that an alkyne molecule contains four hydrogen atoms less than an alkane molecule with the same number of carbon atoms, e.g. C_2H_2 and C_2H_6. The alkynes are more unsaturated than the alkenes.

There are no isomers of the first two alkynes, and two isomers of butyne, C_4H_6.

ethyne (acetylene)

C_2H_2 or $HC{\equiv}CH$ or

$H-C{\equiv}C-H$

propyne

C_3H_4 or $CH_3C{\equiv}CH$ or

```
   H
   |
H—C—C≡C—H
   |
   H
```

$CH_3CH_2C{\equiv}CH$ or $C_2H_5C{\equiv}CH$

$$H{-}\overset{H}{\underset{H}{\overset{|}{\underset{|}{C^4}}}}{-}\overset{H}{\underset{H}{\overset{|}{\underset{|}{C^3}}}}{-}C^2{\equiv}C^1{-}H$$

but-1-yne

$CH_3C{\equiv}CCH_3$

$$H{-}\overset{H}{\underset{H}{\overset{|}{\underset{|}{C^4}}}}{-}C^3{\equiv}C^2{-}\overset{H}{\underset{H}{\overset{|}{\underset{|}{C^1}}}}{-}H$$

but-2-yne

All four atoms of an ethyne molecule are in a straight line and the linear molecule is symmetrical. Make models of the above molecules.

The lengths in nanometres (10^{-9} m) between carbon–carbon single, double and triple bonds are:

$$C{-}C\ 0.154; \quad C{=}C\ 0.134; \quad C{\equiv}C\ 0.120$$

The triple bond is the shortest and most reactive.

Laboratory preparation of ethyne

1. *Action of water on calcium dicarbide* (*calcium ethynediide*). Cover the bottom of a flask with sand, which protects the glass from heat developed during the subsequent reaction, and add small lumps of calcium dicarbide. Fit the flask with a dropping funnel and delivery tube, leading to a gas-jar inverted in a trough of water. Add water drop by drop from the funnel to the dicarbide. Collect the ethyne over water.

$$CaC_2(s) + 2H_2O \rightarrow Ca(OH)_2(aq) + C_2H_2(g)$$

The gas always contains phosphine and hydrogen sulphide, PH_3 and H_2S, as impurities; they are formed from calcium phosphide, Ca_3P_2, and calcium sulphide, CaS, which are always present in ordinary calcium dicarbide.

2. *Elimination of hydrogen halide from a dihalogenoalkane.* Mix alcoholic potassium hydroxide with 1,2-dibromoethane or 1,1-dibromoethane in a flask. Heat the mixture and collect the ethyne, which is pure, over water.

$$\begin{array}{c}CH_2Br\\|\\CH_2Br\end{array} \text{ or } \begin{array}{c}CH_3\\|\\CHBr_2\end{array} \xrightarrow[-2HBr]{KOH(alc)} \begin{array}{c}CH\\|||\\CH\end{array}$$

The reaction takes place in two stages, and the intermediate product is bromoethene:

$$\begin{array}{c} CH_2Br \\ | \\ CH_2Br \end{array} \xrightarrow[-HBr]{KOH(alc)} \begin{array}{c} CH_2 \\ \| \\ CHBr \end{array} \xrightarrow[-HBr]{KOH(alc)} \begin{array}{c} CH \\ ||| \\ CH \end{array}$$

(In organic chemistry, it is common to represent reactions as in the two examples above instead of by balanced equations. These show the reagent and the change involved.) The balanced equation for the above reactions is:

$$C_2H_4Br_2(l) + 2KOH(alc) \rightarrow C_2H_2(g) + 2KBr + 2H_2O$$

Tests on ethyne

1-3. Do the combustion, bromine and potassium manganate(VII) tests as for methane, p. 20. Remember that a mixture of ethyne and air explodes violently if ignited.

4. *Diamminecopper(I)*, $[Cu(NH_3)_2]^+$. Add a small spoonful of copper(II) carbonate to concentrated hydrochloric acid in a test-tube. Add copper turnings and boil gently for 5 minutes. Pour the solution into a second tube half-full of distilled water.

$$CuCl_2(aq) + Cu(s) \rightarrow \underset{\text{copper(I) chloride}}{2CuCl(s)} \quad \text{(white)}$$

Let the precipitate settle and then pour away the liquid. Add concentrated ammonia solution carefully to the white solid until it dissolves to form diamminecopper(I), $[Cu(NH_3)_2]^+$.

Pour this solution into a jar of ethyne. Observe any reaction. (The solid produced is explosive when dry. Destroy it by adding to warm dilute nitric acid.)

5. *Diamminesilver*, $[Ag(NH_3)_2]^+$. To about 2 cm^3 of silver nitrate solution in a test-tube add dilute ammonia solution. A brown precipitate of silver oxide forms at first and then reacts to produce a clear solution called diamminesilver, $[Ag(NH_3)_2]^+$.

$$2Ag^+(aq) + 2OH^-(aq) \rightarrow Ag_2O(s) + H_2O$$

$$Ag^+(aq) + 2NH_3(aq) \rightarrow \underset{\text{diamminesilver(I) ion}}{[Ag(NH_3)_2]^+(aq)}$$

Pour this solution into a jar of ethyne. Observe any reaction. Destroy the precipitate by adding it to warm dilute nitric acid.

6. *Solubility in propanone (acetone).* Invert a boiling tube full of ethyne in a small dish of propanone. Note if the gas is slightly or readily soluble in this solvent.

Structure and shape of ethyne molecules

A carbon atom (2.4) has four electrons available for bonds. One of these electrons bonds with a hydrogen atom and the other bonds with the second carbon atom. All four nuclei in an ethyne molecule are in a straight line and the C—H and C—C bonds are σ-bonds.

Each carbon atom also has two more electrons for forming bonds. Their orbitals are at right angles to each other and to the C—C σ-bond. The electrons overlap to form two π-bonds, each of which can be represented by two charge clouds, Fig. 4.1, in planes at right angles. Therefore the triple bond in ethyne consists of one σ-bond and two π-bonds. Make a ball and spring model of the molecule.

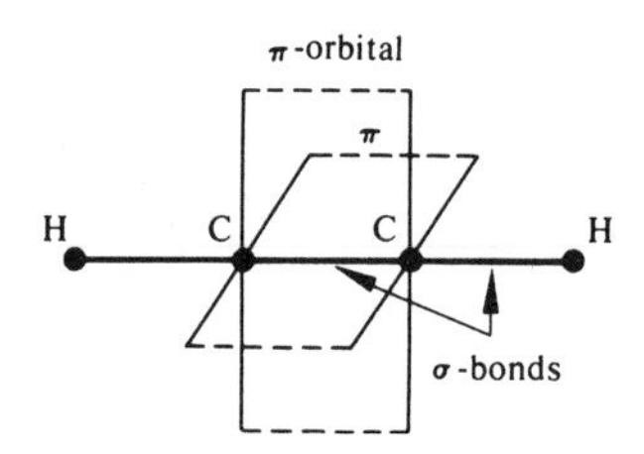

Fig. 4.1. Bonds in an ethyne molecule

The π-electrons of the π-bonds in ethyne (as in ethene and benzene (p. 75)) are around the axis joining the two carbon nuclei (the internuclear axis). The electrons are easy available to electrophiles and therefore the bonds cause the molecule to be very reactive.

Properties of ethyne

Physical properties. Ethyne is a colourless gas. It has a sweetish smell when pure but an unpleasant garlic-like small when impure. It is only slightly soluble in water but very soluble in propanone (acetone). The gas is readily liquefied by pressure but the compressed gas and liquid are dangerously explosive and, unlike oxygen and hydrogen, cannot be stored safely in steel cylinders. Acetylene cylinders contain the gas dissolved in propanone spread on porous silicon(IV) oxide and under a pressure of about 12 atm.

Boiling points of ethyne, propyne and butyne are 190, 250 and 282 K respectively.

Chemical properties

1. *Combustion.* Ethyne burns in air or oxygen with a luminous flame, which is very smoky (ethyne contains 92.3 per cent by mass of carbon):

$$2C_2H_2 + 5O_2 \rightarrow 4CO_2(g) + 2H_2O(g), \text{ and also}$$

$$2C_2H_2 + 3O_2 \rightarrow 4CO + 2H_2O$$

$$2C_2H_2 + O_2 \rightarrow 4C(s) + 2H_2O$$

Ethyne and air form an explosive mixture. Ethyne reacts explosively with chlorine, forming carbon:

$$C_2H_2(g) + Cl_2(g) \rightarrow 2HCl(g) + 2C(s)$$

2. *Addition reactions of ethyne.* An ethyne molecule can combine with two univalent atoms or groups (X and Y) to form an unsaturated molecule with a double bond and then with two more univalent atoms or groups to form a saturated compound:

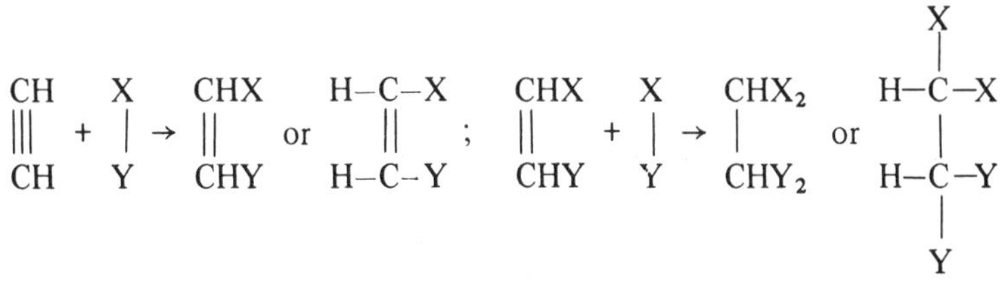

$$\begin{matrix} CH \\ ||| \\ CH \end{matrix} + \begin{matrix} X \\ | \\ Y \end{matrix} \rightarrow \begin{matrix} CHX \\ || \\ CHY \end{matrix} \text{ or } \begin{matrix} H-C-X \\ || \\ H-C-Y \end{matrix} \;;\; \begin{matrix} CHX \\ || \\ CHY \end{matrix} + \begin{matrix} X \\ | \\ Y \end{matrix} \rightarrow \begin{matrix} CHX_2 \\ | \\ CHY_2 \end{matrix} \text{ or } \begin{matrix} X \\ | \\ H-C-X \\ | \\ H-C-Y \\ | \\ Y \end{matrix}$$

The two X groups combine with one carbon atom and two Y groups combine with the other. Symmetrical addition compounds with the formula CHXYCHXY are not formed, that is, one X and one Y group do not attach themselves to the same carbon atom. X and Y may, of course, be the same, e.g. H, Cl or Br, when hydrogen, chlorine or bromine are the addition reagents.

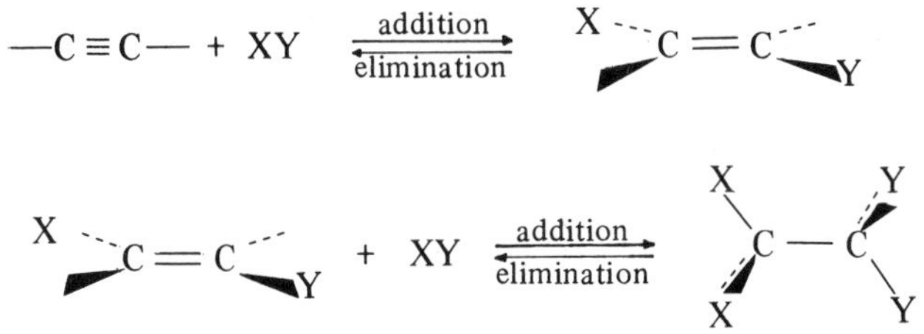

$$-C \equiv C- + XY \underset{\text{elimination}}{\overset{\text{addition}}{\rightleftharpoons}} XC=CY$$

$$XC=CY + XY \underset{\text{elimination}}{\overset{\text{addition}}{\rightleftharpoons}} X_2C-CY_2$$

(a) *Hydrogen.* Hydrogen gas and ethyne combine in the presence of a heated nickel, platinum or palladium catalyst.

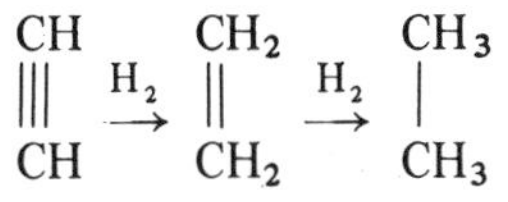

$$\begin{matrix} CH \\ ||| \\ CH \end{matrix} \xrightarrow{H_2} \begin{matrix} CH_2 \\ || \\ CH_2 \end{matrix} \xrightarrow{H_2} \begin{matrix} CH_3 \\ | \\ CH_3 \end{matrix}$$

(b) *Halogens.* Chlorine and ethyne combine when passed into an inert solvent containing antimony pentachloride as catalyst. The reaction is carried out commercially.

$$\begin{matrix} CH \\ ||| \\ CH \end{matrix} + \begin{matrix} Cl \\ | \\ Cl \end{matrix} \rightarrow \begin{matrix} CHCl \\ || \\ CHCl \end{matrix} \quad \text{then} \quad \begin{matrix} CHCl \\ || \\ CHCl \end{matrix} + \begin{matrix} Cl \\ | \\ Cl \end{matrix} \rightarrow \begin{matrix} CHCl_2 \\ | \\ CHCl_2 \end{matrix}$$

The products are 1,2-dichloroethene and 1,1,2,2-tetrachloroethane.

Bromine vapour, liquid bromine and bromine water react with ethyne at room temperature, and they are decolorized by excess ethyne.

$$C_2H_2 \xrightarrow{Br_2} C_2H_2Br_2 \xrightarrow{Br_2} C_2H_2Br_4$$

(c) *Hydrogen halides.* Hydrogen iodide reacts readily at room temperature, hydrogen bromide reacts at about 370 K, and hydrogen chloride reacts only in the presence of a catalyst, which is mercury(II) chloride on activated charcoal:

$$\begin{matrix} CH \\ ||| \\ CH \end{matrix} + \begin{matrix} H \\ | \\ I \end{matrix} \rightarrow \begin{matrix} CH_2 \\ || \\ CHI \end{matrix} \quad \text{then} \quad \begin{matrix} CH_2 \\ || \\ CHI \end{matrix} + \begin{matrix} H \\ | \\ I \end{matrix} \rightarrow \begin{matrix} CH_3 \\ | \\ CHI_2 \end{matrix}$$

The products are iodoethene and 1,1-diiodoethane respectively.

$$\begin{matrix} CH \\ ||| \\ CH \end{matrix} + \begin{matrix} H \\ | \\ Cl \end{matrix} \xrightarrow[\text{charcoal}]{HgCl_2} \begin{matrix} CH_2 \\ || \\ CHCl \end{matrix}$$

chloroethene

This reaction is used to manufacture chloroethene which is then converted to its useful polymer poly(chloroethene) (p. 54).

(d) *Water.* Water combines with ethyne in the presence of a catalyst, which is mercury(II) sulphate in warm 30 per cent sulphuric acid:

$$HC{\equiv}CH \xrightarrow[H_2SO_4]{H_2O,\ Hg^{2+}} (H_2C{=}CHOH) \rightarrow CH_3CHO$$

ethanal

The equation shows a possible unstable intermediate product.

(e) *Potassium manganate(VII).* The pink solution of acidified manganate is decolorized by excess ethyne, which is oxidized:

$$\begin{matrix} CH \\ ||| \\ CH \end{matrix} + 4[O]\ [\text{from } KMnO_4] \rightarrow \begin{matrix} COOH \\ | \\ COOH \end{matrix}$$

ethanedioic acid

3. *Precipitation of ethynediides.* Red dicopper(I) ethynediide and white disilver(I) ethynediide are precipitated when ethyne is passed into diammine-copper(I) chloride and diamminesilver nitrate respectively:

$$C_2H_2 + 2CuCl \rightarrow 2HCl + Cu_2C_2(s) \quad \text{or} \quad Cu{-}C{\equiv}C{-}Cu \text{ (red)}$$

$$C_2H_2(g) + 2[Cu(NH_3)_2]^+(aq) \rightarrow 2NH_4^+(aq) + 2NH_3(aq) + Cu_2C_2(s)$$

Both solids are dangerously explosive when dry.

4. *Polymerization.* Ethyne polymerizes to its trimer, benzene, when passed through a red-hot tube:

$$3C_2H_2(g) \rightarrow C_6H_6(g)$$

A benzene molecule contains a ring of six carbon atoms; the polymerization is an example of the conversion of a chain hydrocarbon to a ring or cyclic hydrocarbon.

In the presence of certain catalysts, ethyne forms its *dimer*, which is used to make a synthetic rubber:

$$2CH{\equiv}CH \rightarrow CH{\equiv}C{-}CH{=}CH_2$$

Uses of ethyne

The temperature of the oxyacetylene flame is more than 2200 K, and the flame is used to cut and to weld metals.

A little ethanal is made by the addition of water to ethyne (p. 64), but the process is now of minor importance. Addition of hydrogen chloride to ethyne yields chloroethene, used to manufacture poly(chloroethene) (polyvinyl chloride), an important plastic.

Manufacture of ethyne

Most ethyne is obtained from other hydrocarbons. It is produced when certain fractions from petroleum are cracked. Natural gas is also cracked by heating to 1800 K for 0.01 s and then cooling rapidly:

$$2CH_4 \rightarrow C_2H_2 + 3H_2 \quad \text{(dehydrogenation)}$$

Partial combustion of natural gas at high temperature is also used:

$$4CH_4 + 3O_2 \rightarrow 2C_2H_2 + 6H_2O$$

Addition of bromine to ethyne (electrophilic reaction)

Bromine adds on to ethyne by a mechanism similar to that with ethene (p. 49), except that two products are formed:

$$HC{\equiv}CH \rightarrow BrHC{=}CHBr \rightarrow Br_2HC{-}CHBr_2$$

In the first product, the two bromine atoms are on opposite sides of the planar molecule. The two CHBr groups cannot rotate around the double bond C=C, which is rigid and differs in this respect from a single bond C—C. The mechanism can be represented as below, in which a curved arrow represents a shift of two electrons (p. 49):

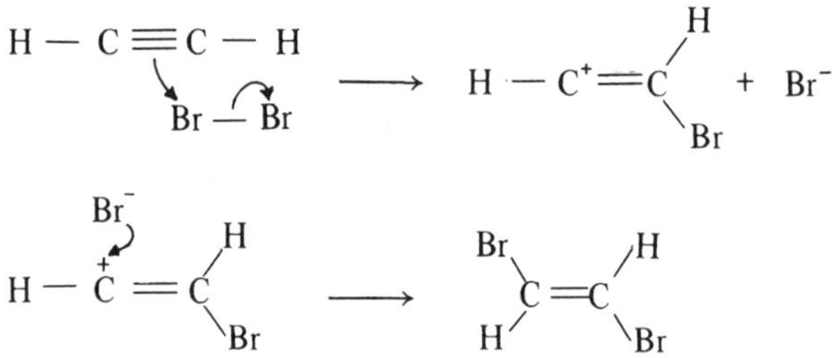

$$H{-}C{\equiv}C{-}H \;\; Br{-}Br \longrightarrow H{-}C^{+}{=}C(H)(Br) + Br^{-}$$

$$Br^{-} + H{-}\overset{+}{C}{=}C(H)(Br) \longrightarrow (Br)(H)C{=}C(H)(Br)$$

The product is called *trans*-1,2-dibromoethene because the bromine atoms are on opposite sides. (The compound with the bromine atoms on the same side is *cis*-1,2-dirbromoethene.)

SUMMARY

Alkynes

C_nH_{2n-2}. Functional group: C≡C. Triple bond is σ-bond and two π-bonds. Electrophilic reagents attack the weak π-bonds.

Preparation

1. Calcium dicarbide and cold water → C_2H_2 (ethyne)

2. Dihalogenoalkane and KOH(alc).

$$CH_2BrCH_2Br \text{ or } CH_3CHBr \rightarrow C_2H_2$$

Properties
(of C_2H_2 or CH≡CH, ethyne)

Combustion: Air Chlorine	 $CO_2 + H_2O$. Sooty, luminous flame. Explosive $C + H_2O$. Sooty flame. Explosive
Addition:	
H_2/Ni or Pt	CH_2CH_2 then CH_3CH_3 ethene ethane
Cl_2(g)/catalyst	CHCl=CHCl then $CHCl_2CHCl_2$ 1,2-dichloroethene 1,1,2,2-tetrachloroethane Industrial process
Br_2(g or l or CCl_4)	CHBr=CHBr, then $CHBr_2CHBr_2$
HI(g), 20 °C	CH_2CHI then CH_3CHI_2 iodoethene 1,1-diiodoethane
HBr(g), 100 °C	CH_2CHBr then CH_3CHBr_2
HCl(g)/catalyst	CH_2=CHCl Industrial process for PVC chloroethene
H_2O/Hg^{2+} and H_2SO_4	CH_3CHO ethanal
$KMnO_4/H^+$	Decolorized COOHCOOH ethanedioic acid
$[Cu(NH_3)_2]^+$(aq)	Cu_2C_2(s). Red dicopper(I) ethynediide
$[Ag(NH_3)_2]^+$(aq)	Ag_2C_2(s). White disilver(I) ethynediide
Heat	Polymerizes to C_4H_4 then C_6H_6 benzene

Manufacture
Natural gas cracked at very high temperatures for 0.01 s. Also partial oxidation of natural gas by air at high temperatures.

Mechanism of addition reactions
Similar to those with ethene.

QUESTIONS

1. Outline methods by which (a) ethyne, and (b) propyne may be prepared in the laboratory.

2. Mention briefly three reactions in which ethyne acts as an unsaturated compound, and give equations. Give two other reactions of ethyne which do not involve addition.

3. Describe chemical tests by which you would distinguish between (a) ethyne and ethene, and (b) ethene and ethane.

4. Discuss the mechanism of the addition reactions between ethyne and liquid bromine.

5. Explain why a carbon–carbon double bond is much more reactive than a carbon–carbon single bond. Three types of reaction are ***addition, substitution*** and ***elimination.*** By giving a suitable example of each of these reaction types explain briefly the meaning of the terms.

6. 8 cm^3 of a hydrocarbon were mixed with 78 cm^3 of oxygen and the mixture was exploded. The volume of the products was 62 cm^3 (all volumes measured at room temperature and pressure). Aqueous alkali absorbed 40 cm^3 of carbon dioxide from the products. Show that the formula of the hydrocarbon is probably C_5H_8. Write possible structural formulae for the hydrocarbon.

Chapter 5

BENZENE AND METHYLBENZENE

Molecules of alkanes, alkenes and alkynes have carbon atoms in chains which are either unbranched or branched. Molecules of cyclic hydrocarbons have carbon atoms in a ring, and one or more carbon atoms may be attached to the ring.

Cycloalkanes are saturated hydrocarbons of the general formula C_nH_{2n} where n is 3 or more and the carbon atoms are in a ring. A cycloalkane is named by adding *cyclo* to the name of the alkane with the same number of carbon atoms per molecule. A cycloalkane molecule contains two atoms less than the corresponding alkane molecule, e.g. hexane is C_6H_{14} and cyclohexane is:

C_6H_{12} or or or

Cycloalkenes are unsaturated compounds of the general formula C_nH_{2n-2}. Each molecule contains one double bond, C=C. The formula of hexene is C_6H_{12}, and that of cyclohexene is:

C_6H_{10} or or or

Benzene and methylbenzene (toluene) are two cyclic hydrocarbons known as *arenes* and sometimes called *aromatic hydrocarbons* because of the pleasant odour of some benzene derivatives. The formula of benzene (p. 73) is:

C_6H_6 or [structural formula] or [hexagon with circle]

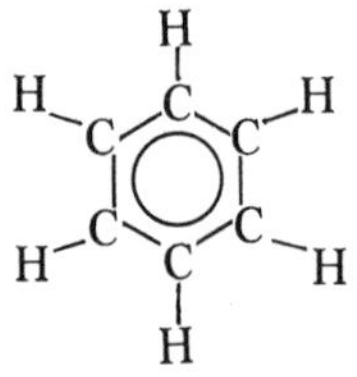

The formula of methylbenzene, which has a methyl *side chain*, is:

C_7H_8 or $C_6H_5CH_3$ or [structural formula] or [hexagon with circle, CH_3]

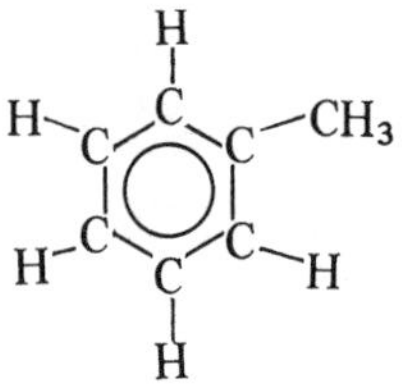

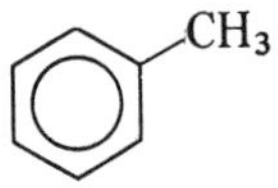

Make models of molecules of the above cyclic hydrocarbons.

To study some properties of four cyclic hydrocarbons

The hydrocarbons and their boiling points in kelvins are: cyclohexane, 354; cyclohexene, 356; benzene, 353; methylbenzene, 384.

Benzene liquid and vapour are poisonous and therefore should be handled only in a fume cupboard.

1. *Combustion.* Add 5 drops of the liquids separately to four hard-glass watch-glasses in a fume cupboard. Remove the bottles of the liquids to a safe place. Ignite each liquid in turn and observe the flames.

2. *Bromine.* Add about 2 cm^3 of the liquids separately to four test-tubes in a fume cupboard. Use a teat pipette to add 4 drops of liquid bromine to each tube, and shake well. Observe if the bromine is decolorized in any of the tubes and if the change is rapid or slow.

Add iron filings to the test-tube containing bromine and benzene.

3. *Potassium manganate(VII).* Add about 1 cm^3 of the liquids separately to test-tubes. Add 4 drops of very dilute neutral potassium manganate solution to each tube, and shake well. Observe if the pink manganate is decolorized. If necessary, warm the solutions gently but be careful that they do not ignite.

Repeat the test with potassium manganate acidified with dilute sulphuric acid.

4. *Sulphuric acid.* Add about 2 cm^3 of concentrated sulphuric acid to four test-tubes in a rack (not in your hand). Add about 10 drops of the hydrocarbons separately to the tubes. Note which hydrocarbons are miscible with the acid and which also seem to react.

Add about 2 cm^3 of the fuming acid to 1 cm^3 of benzene, and be prepared for a vigorous reaction. Repeat the test with methylbenzene.

5. *Nitric acid.* To a mixture of 5 cm^3 of concentrated sulphuric acid and 1 cm^3 of concentrated nitric acid, add 1 cm^3 of benzene drop by drop and shake continuously. Note any colour change. Pour the product carefully into a beaker of cold water. Unchanged benzene floats; observe any oily drops which sink.

Repeat the test with methylbenzene.

Repeat the whole test, with benzene and then methylbenzene, using a mixture of 1 cm^3 of concentrated sulphuric acid and 5 cm^3 of concentrated nitric acid. Boil the mixture of acid and hydrocarbon carefully for 1 minute before adding to cold water.

BENZENE

Commercial preparation. Most benzene (and methylbenzene) is obtained from petroleum. Certain fractions are subjected to carefully controlled processes called *catalytic reforming* and *platforming.* In the former, petroleum naphtha is passed over a catalyst such as oxides of aluminium and molybdenum, and in the platforming process the naphtha vapour is passed over a catalyst of platinum on aluminium oxide at about 770 K and 20 atm pressure. The products are cooled; they form gaseous compounds, mainly hydrogen, and liquid hydrocarbons such as benzene, methylbenzene, other aromatic hydrocarbons and alkanes. Pure benzene and methylbenzene are separated by fractionation and by extraction with solvents which do not dissolve alkanes. (Sometimes the liquid hydrocarbons are added to petrol without further treatment.)

Three changes occur during reforming or platforming (note that the number of carbon atoms per molecule does not change; it does during cracking):

(a) alkanes with unbranched chains form ring compounds which then lose more hydrogen; e.g. hexane forms cyclohexane and then benzene:

$$\mathrm{CH_3CH_2CH_2CH_2CH_2CH_3} \xrightarrow{\mathrm{Pt/Al_2O_3}} \text{[cyclohexane]} + \mathrm{H_2} \xrightarrow{\mathrm{Pt/Al_2O_3}} \text{[benzene]} + \mathrm{4H_2}$$

(b) cycloalkanes isomerize to a cyclohexane compound which then dehydrogenates:

$$\underset{\text{methylcyclopentane}}{C_5H_9CH_3} \longrightarrow \underset{\text{cyclohexane}}{C_6H_{12}} \longrightarrow C_6H_6 + 3H_2$$

(c) dehydrogenation of cycloalkanes:

$$\underset{\text{cyclohexane}}{C_6H_{12}} \longrightarrow C_6H_6 + 3H_2$$

$$\underset{\text{methylcyclohexane}}{C_6H_{11}CH_3} \longrightarrow C_6H_5CH_3 + 3H_2$$

Some benzene and methylbenzene are still obtained from coal but the process is of minor importance. Coal is heated in the absence of air (destructive distillation). Two of the products are coal tar and coal gas and both contain benzene and methylbenzene. The gas is passed through an oil which retains the arenes. The tar is fractionated and the two arenes are in the fraction boiling below 443 K and known as *light oil* because it is less dense than water.

The two liquids containing the arenes are mixed and then treated with sodium hydroxide to remove acids and with sulphuric acid to remove bases. The liquid is then fractionated into five fractions. The second fraction contains most of the benzene and the third contains the methylbenzene. Further fractional distillation yields the pure compounds. Benzene is purified commercially by crystallization (it freezes at 278.7 K).

Laboratory preparation of benzene

Benzene is seldom prepared because it can be bought cheaply. Processes which can be used to prepare it are:

1. Heat benzoic acid or its sodium, potassium, calcium or barium salt with sodalime:

$$C_6H_5COONa(s) + NaOH(s) \rightarrow Na_2CO_3(s) + C_6H_6(g)$$

compare $$CH_3COONa(s) + NaOH(s) \rightarrow Na_2CO_3(s) + CH_4(g)$$

2. Pass phenol vapour over hot zinc dust:

$$C_6H_5OH(g) + Zn(s) \rightarrow ZnO(s) + C_6H_6(g)$$

Molecular formula of benzene

Benzene consists of 92.3 per cent of carbon and 7.7 per cent of hydrogen by mass. The ratio of carbon to hydrogen by atoms is therefore $92.3/12 : 7.7/1 = 7.7 : 7.7 = 1 : 1$. The empirical formula is therefore CH and the molecular formula is $(CH)_n$. The relative molecular mass is $(12 + 1)n = 13n$, and experiments show that it is 78. Therefore $13n = 78$ and $n = 6$. The molecular formula of benzene is C_6H_6.

It is possible to write a chain formula for benzene, e.g. $CH{\equiv}CCH_2CH_2C{\equiv}CH$. This compound would be very unsaturated, like ethyne, and this is not true of benzene. Also there is only one monosubstituted benzene C_6H_5X but the chain formula permits two isomers.

Kekulé formula for benzene

In 1865, Kekulé suggested that the carbon atoms in a benzene molecule are in a regular hexagonal ring, now called a *benzene ring* or *aromatic ring*. The Kekulé formula is:

CH, HC, CH, HC, CH, CH or H, C, H, C, C, H, C, C, H, C, H or

It is usual to omit the carbon and hydrogen atoms when writing the formula.

The formula readily accounts for only one monosubstituted benzene. However, there should be two 1,2-disubstituted benzenes, $C_6H_4X_2$, because the X groups could be joined to carbon atoms separated by a single or a double bond:

X X and X X

Similarly there should be two 1,2-compounds C_6H_4XY. Since 1,2-disubstituted benzenes do not have two isomers, Kekulé assumed that the double bonds were mobile and that benzene is an equilibrium mixture:

$$\rightleftharpoons$$

Objections to Kekulé formula for benzene

The Kekulé formula shows three alkene double bonds, C=C. Benzene should therefore be very unsaturated. It forms addition compounds with hydrogen, chlorine, bromine and ozone, but not with hydrochloric acid, sulphuric acid and potassium manganate(VII). Also, benzene can form substitution products with chlorine, bromine, sulphuric acid, nitric acid and halogenoalkanes. Therefore benzene is not unsaturated in the same way as alkenes and its double bonds are not true alkene double bonds.

Symmetry of the benzene ring. X-rays are diffracted by a benzene crystal, and the diffraction or scattering is caused by electrons. The diffraction is greater where the electron density is greater, that is, near the carbon nuclei because the hydrogen nuclei are smaller and have little effect. The electron densities at various points of a benzene molecule have been determined, and from them the shape of the molecule and the bond lengths (distances between nuclei) have been calculated. The carbon–carbon bond lengths are equal and they are not alternately double and single bonds as in the Kekulé formula.

The bond lengths in nanometres are:

$$\text{C—C } 0.154 \qquad \text{C=C } 0.133 \qquad \text{C—C in benzene } 0.139$$

The carbon–carbon bonds in benzene are intermediate between true single and true double bonds and they are neither one nor the other.

Heat of formation. The bonds in a Kekulé benzene molecule and their bond energies in kilojoules per mole are:

$$\text{six C—H } (413) \qquad \text{three C—C } (346) \qquad \text{three C=C } (611)$$

The total bond energy of benzene should be:

$$(6 \times 413) + (3 \times 346) + (3 \times 611)\ \text{kJ mol}^{-1}$$
$$= 2478 + 1038 + 1833$$
$$= 5349\ \text{kJ mol}^{-1}$$

The following equation should be correct:

$$6C(g) + 6H(g) \rightarrow C_6H_6(g) \qquad \Delta H = -5349\ \text{kJ mol}^{-1}$$

However the experimental value for the heat of formation of benzene gas is -5514 kJ mol^{-1}. The figures show that benzene contains 165 kJ mol^{-1} *less* energy than that required by the Kekulé structure.

Heat of hydrogenation. Heat is evolved during the hydrogenation of cyclohexene, whose molecule contains one double bond C=C.

$$C_6H_{10}(g) + H_2(g) \rightarrow C_6H_{12}(g) \qquad \Delta H = -120\ \text{kJ mol}^{-1}$$

If a benzene molecule contains three double bonds C=C, the heat of hydrogenation of benzene should be $3 \times -120 = -360$ kJ mol^{-1}, but the experimental value is only -208 kJ mol^{-1}:

$$C_6H_6(g) + 3H_2(g) \rightarrow C_6H_{12}(g) \qquad \Delta H = -208\ \text{kJ mol}^{-1}$$

Therefore less heat is evolved during the hydrogenation of benzene than that required by the Kekulé formula. The figures mean that benzene has $360 - 208 = 152\ \text{kJ mol}^{-1}$ *less* energy than expected.

Modern view of benzene structure

Each carbon atom (2.4) has four electrons available for forming bonds. Three of these electrons join a carbon atom to two other carbon atoms and one hydrogen atom by σ-bonds. The twelve atoms are all in one plane. The six other electrons are not between a particular pair of carbon atoms. They are free to move between all the six carbon atoms in the ring. These six electrons are said to be *delocalized*. Their electron charge cloud lies above and below the plane of the ring, forming a π-orbital around the ring (Fig. 5.1(b)).

(Compounds which contain a *conjugated system*, of alternating single and double bonds —C=C—C=C—, also have delocalized electrons and do not contain true single and double bonds. Buta-1,3-diene, $CH_2{=}CH{-}CH{=}CH_2$, is one example.)

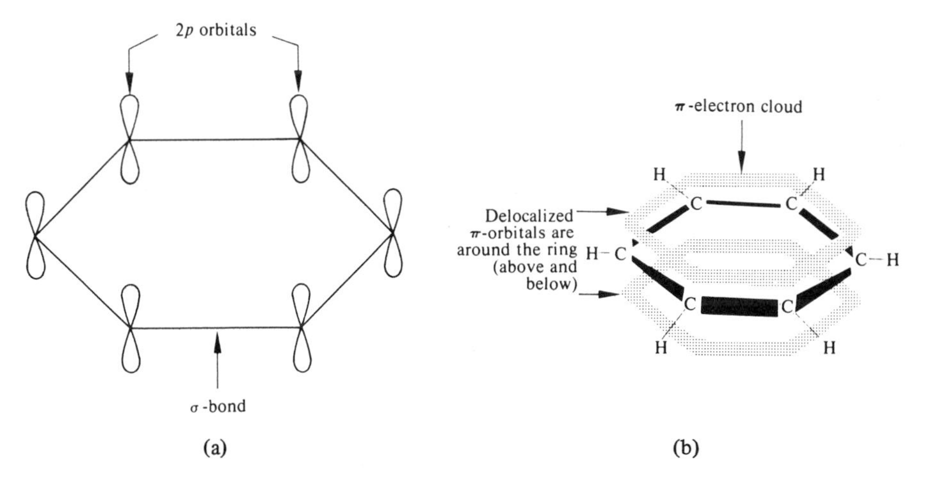

Fig. 5.1. Structure of a benzene ring

Properties of benzene

Physical properties. It is a colourless inflammable liquid with a characteristic smell, and it is poisonous. It boils at 353.25 K, freezes at 278.7 K, and its density is $0.88\ \text{g cm}^{-3}$ at 298 K. It is practically insoluble in water. It is a good solvent for iodine, phosphorus, fats, oils, resins, and many organic compounds.

Chemical properties. Benzene is stable. It is not affected by alkalis, hydrochloric acid, and oxidizing agents.

1. *Combustion.* Benzene burns with a smoky luminous flame because of its high carbon content (92.3 per cent by mass):

$$2C_6H_6 + 15O_2 \rightarrow 12CO_2 + 6H_2O, \text{ and also}$$

$$2C_6H_6 + 9O_2 \rightarrow 12CO + 6H_2O$$

$$2C_6H_6 + 3O_2 \rightarrow 12C + 6H_2O$$

2. *Addition reactions of benzene.* Benzene should be unsaturated and its molecule should contain three double bonds. However, there are few addition reactions of benzene, and therefore it is not unsaturated in the same way as alkenes and cycloalkenes.

(a) *Hydrogen.* Benzene vapour and hydrogen combine under pressure and with a catalyst, which can be nickel at 470 K or finely divided platinum known as platinum black. The hydrogen adds in three stages; however, the two intermediate products cannot be obtained by direct hydrogenation.

$$\underset{\text{benzene}}{C_6H_6(g)} \longrightarrow \underset{\text{cyclohexadiene}}{C_6H_8} \longrightarrow \underset{\text{cyclohexene}}{C_6H_{10}} \longrightarrow \underset{\text{cyclohexane}}{C_6H_{12}}$$

$$\text{benzene ring} \xrightarrow[\text{470 K, pressure}]{H_2,\ Ni} \text{cyclohexane}$$

cyclohexane

(b) *Chlorine.* Chlorine combines when it is bubbled through boiling benzene in bright sunlight. No catalyst is necessary. The high temperature and light favour the formation of chlorine atoms which take part in the addition reactions:

$$C_6H_6 \xrightarrow{Cl_2} C_6H_6Cl_2 \xrightarrow{Cl_2} C_6H_6Cl_4 \xrightarrow{Cl_2} C_6H_6Cl_6$$

$$\text{benzene ring} + 3Cl_2 \xrightarrow[\text{bright light}]{\text{boiling}} \text{cyclohexane ring with six Cl}$$

The final product is hexachlorocyclohexane. It is sold under the name *Gammexane* as an insecticide used on poultry, cattle and sheep.

(c) *Ozone.* Ozone adds slowly to form a solid triozonide. The benzene molecules react as though each contains three double bonds, C=C. Water hydrolyses the triozonide to break the molecules at these 'double bonds' and yield the dialdehyde called ethanedial:

$$C_6H_6 \xrightarrow{O_3} C_6H_6(O_3)_3 \xrightarrow{H_2O} \begin{array}{c} CHO \\ | \\ CHO \end{array} + H_2O_2$$

3. *Substitution reactions of benzene.* A univalent atom or group can displace a hydrogen atom and form a compound C_6H_5X in which X may be Cl, Br, SO_2OH, NO_2, CH_3 or other alkyl group, or $COCH_3$. The products are mono-substituted benzene derivatives or *phenyl* compounds. The phenyl group is C_6H_5—; it is an *aryl* group, the aromatic equivalent of an alkyl group.

(a) *Halogenation.* Chlorine and bromine substitute in the absence of sunlight and with a catalyst such as anhydrous iron(III) chloride or bromide, anhydrous aluminium chloride, or iodide. These catalysts are known as *halogen carriers.* The reaction between iodine and benzene is negligible, whereas that between fluorine and benzene is too vigorous to control.

$$C_6H_6 \xrightarrow[AlCl_3 \text{ or } FeCl_3]{Cl_2} C_6H_5Cl \;;\quad C_6H_6 + Br_2 \xrightarrow[\text{catalyst}]{FeBr_3} C_6H_5Br + HBr$$

The catalyst polarizes the halogen molecules and finally favours fission (heterolysis). The reagents which attack the benzene ring are the electrophiles Cl^+ and Br^+:

$$Cl{-}Cl \xrightarrow{\text{catalyst}} Cl{:}^- + Cl^+ \text{ (heterolytic fission)}$$

$$Br_2 + FeBr_3 \rightarrow \overset{\delta+}{Br}{-}\overset{\delta-}{Br}\cdot FeBr_3, \text{ and then}$$

$$C_6H_6 + \overset{\delta+}{Br}{-}\overset{\delta-}{Br}\cdot FeBr_3 \rightarrow [C_6H_6Br]^+ + FeBr_4^- \rightarrow C_6H_5Br + HBr + FeBr_3$$

(b) *Sulphuric acid.* Benzene reacts readily at room temperature with the fuming acid but only with the concentrated acid when heated for several hours under reflux (see Fig. 5.2). The acid formed is a crystalline solid, readily soluble in water:

$$C_6H_6 + HOSO_2OH \rightarrow H_2O + \underset{\text{benzenesulphonic acid}}{C_6H_5SO_2OH}$$

$$C_6H_6 + H_2SO_4 \rightarrow C_6H_5SO_2OH$$

The disulphonic acid is formed from benzene and fuming sulphuric acid at 520 K:

$$C_6H_6 \xrightarrow{\text{fuming } H_2SO_4} C_6H_4(SO_3H)_2 \text{ (1,3-)}$$

Reactions in which a sulphonic group, SO_3H or SO_2OH, replaces a hydrogen atom are called *sulphonations.*

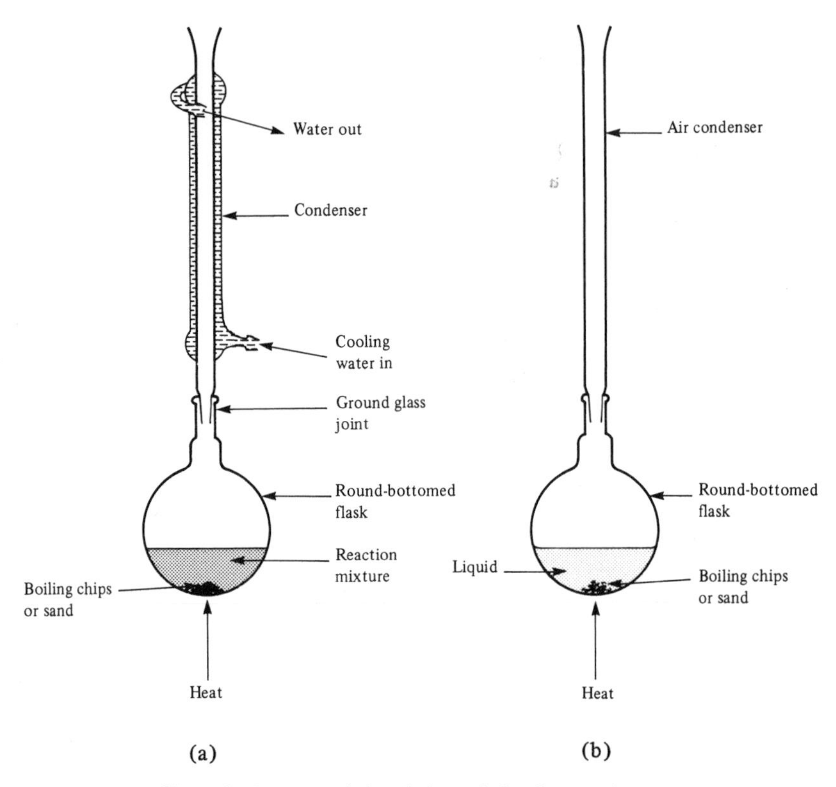

Fig. 5.2. Water-cooled and air-cooled reflux condensers

(c) *Nitric acid.* This acid reacts very slowly with benzene. A mixture of concentrated nitric and sulphuric acids at 320 K nitrates benzene:

$$C_6H_6 \xrightarrow[\substack{H_2SO_4 \\ (-H_2O)}]{HNO_3} C_6H_5-NO_2$$

nitrobenzene

Nitrobenzene is a yellow oily liquid which is denser than water.

Nitration means the substitution of one (or more) hydrogen atoms by the nitro-group, NO_2. The nitrating agent is a mixture of acids and is rarely nitric acid alone.

At temperatures over 320 K, a second nitro group joins to the benzene ring. The product is a yellow solid:

$$C_6H_6 + 2HONO_2(2HNO_3) \rightarrow C_6H_4(NO_2)_2 + 2H_2O$$

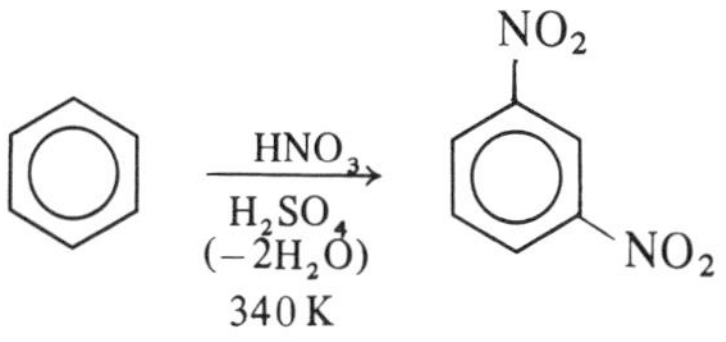

$$C_6H_6 \xrightarrow[\substack{H_2SO_4 \\ (-2H_2O) \\ 340\,K}]{HNO_3} C_6H_4(NO_2)_2$$

1,3-dinitrobenzene

The nitrating mixture contains the electrophile NO_2^+, the nitryl cation, and this attacks the benzene ring. The reaction between the two acids is:

$$HNO_3 + 2H_2SO_4 \rightarrow H_3O^+ + 2HSO_4^- + NO_2^+$$

Nitration and sulphonation are distinctive reactions of arenes and other aromatic compounds.

(d) *Aliphatic halogen compounds (Friedel-Crafts reaction)*. With anhydrous aluminium chloride as a catalyst, dry benzene reacts in the cold with dry halogenoalkanes or acid chlorides to form alkyl or acyl derivatives. Water must be absent because it would hydrolyse the catalyst:

$$C_6H_6 + ClCH_3 \xrightarrow{AlCl_3} HCl + C_6H_5CH_3$$

methylbenzene

then

$$C_6H_5CH_3 + ClCH_3 \xrightarrow{AlCl_3} HCl + C_6H_4(CH_3)_2$$

dimethylbenzene

$$C_6H_6 + ClCOCH_3 \xrightarrow{AlCl_3} HCl + C_6H_5COCH_3$$

phenylethanone

The general equation for the reactions is:

$$C_6H_6 + ClR \xrightarrow{AlCl_3} HCl + C_6H_5R$$

in which R is an alkyl or acyl group, e.g. CH_3 or CH_3CO.

The equations in this section can be written:

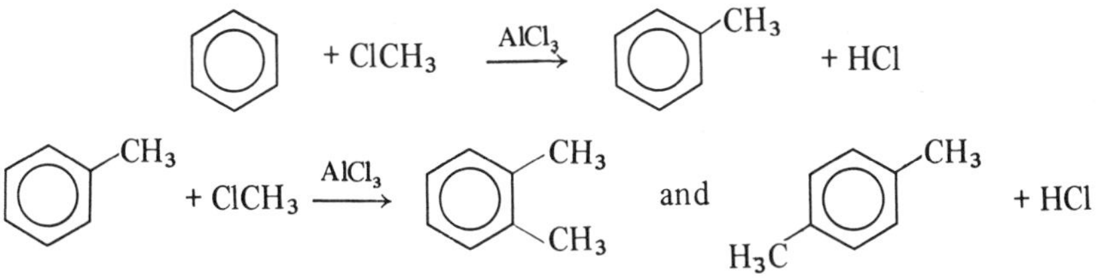

$$C_6H_6 + ClCH_3 \xrightarrow{AlCl_3} C_6H_5CH_3 + HCl$$

$$C_6H_5CH_3 + ClCH_3 \xrightarrow{AlCl_3} \text{1,2-}C_6H_4(CH_3)_2 \text{ and } \text{1,4-}C_6H_4(CH_3)_2 + HCl$$

$$C_6H_6 + ClCOCH_3 \xrightarrow{AlCl_3} C_6H_5COCH_3 + HCl$$

The first equation shows the *alkylation* of benzene to methylbenzene, the second represents the alkylation of methylbenzene to 1,2- and 1,3-dimethylbenzenes, and the third represents the *acylation* of benzene to phenylethanone. Iron(III) bromide is a catalyst used with bromoalkanes, e.g. CH_3Br, C_2H_5Br, etc.

4. *Potassium manganate(VII)*. Under ordinary conditions this reagent has no action on benzene. If benzene contained ordinary double bonds, C=C, the reagent would be expected to add OH groups to these, as with ethene. Boiling potassium manganate(VII) attacks benzene only slowly; the products (ethanedioic acid and propanoic acid, C_2H_5COOH) are produced by breaking the benzene ring. The slowness of the reaction emphasizes the great stability of the benzene ring.

Uses of benzene and methylbenzene

Benzene and methylbenzene are used as solvents. They are added to some petrols. Explosives, dyes, antiseptics and many pharmaceutical drugs are made from them.

Nitration of benzene (electrophilic reaction)

A mixture of concentrated nitric and sulphuric acids contains the nitryl cation, NO_2^+. This electrophile attracts one electron pair from the six π-orbital electrons of benzene and joins NO_2 to one carbon atom by a new σ-bond. The four remaining π-electrons are shared by the other five carbons atoms. The formula of the intermediate addition compound is

$[C_6H_6NO_2]^+$ (ring with H, H, H, H, H and H/NO_2 on one carbon) or (ring with NO_2 and H on one carbon, + inside circular line)

(the circular line represents the limits of movement of four π-electrons)

This compound now loses a proton (*deprotonation*) and the π-electrons of the benzene again occupy their original position. The proton adds to a HSO_4^- ion in the acid mixture, forming sulphuric acid.

$$[C_6H_6NO_2]^+ + HSO_4^- \rightarrow C_6H_5NO_2 + H_2SO_4$$

Halogenation of benzene has a similar mechanism but a halogen carrier catalyst such as $FeCl_3$ or $AlBr_3$ helps to polarize the halogen molecules before electrophilic attack occurs. Note that substitution in benzene has an 'ionic' mechanism whereas substitution in methane has a 'free radical' mechanism. Benzene molecules are electron-rich and act as nucleophiles.

Mechanism of alkylation of benzene

An aluminium atom has three electrons in its outermost orbital and aluminium chloride (or bromide) has three covalent bonds:

$$\begin{array}{c} Cl \\ Cl:\ddot{\underset{\cdot\cdot}{Al}} \\ Cl \end{array} \leftarrow\text{---} \text{ electron-deficient aluminium atom}$$

It readily forms a complex with a halogenoalkane:

$$CH_3Cl + AlCl_3 \rightarrow \overset{+}{C}H_3\overset{-}{A}lCl_4 \text{ (complex carbocation)}$$

This complex then reacts like the nitryl cation above:

$$C_6H_6 + \overset{+}{C}H_3\overset{-}{A}lCl_4 \rightarrow C_6H_6CH_3[AlCl_4] \text{ (slow)}$$

$$C_6H_6CH_3[AlCl_4] \rightarrow C_6H_5CH_3 + HCl + AlCl_3 \text{ (fast)}$$

Names and formulae of substituted benzene compounds

When one or more of the six hydrogen atoms in a benzene molecule are replaced by a monovalent atom or group X, the numbers of possible isomers are:

C_6H_5X 1; $C_6H_3X_3$ 3; C_6HX_5 1

$C_6H_4X_2$ 3; $C_6H_2X_4$ 3; C_6X_6 1

Displayed formulae of the three disubstituted isomers are:

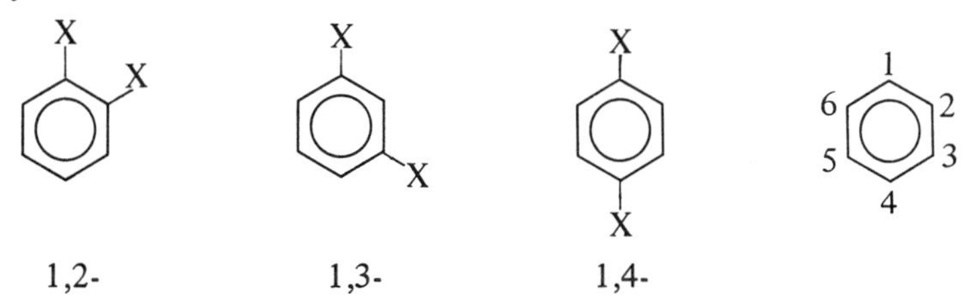

The carbon atoms in a benzene ring are numbered 1 to 6. For example, if X = Cl the three compounds are 1, 2-dichlorobenzene, 1,3-dichlorobenzene and 1,4-dichlorobenzene. Older names are *ortho*-, *meta*- and *para*-dichlorobenzenes or *o*-, *m*- and *p*-compounds.

METHYLBENZENE

Laboratory preparation of methylbenzene

Methylbenzene is a commercial product and is seldom prepared in the laboratory. Processes by which it can be prepared are:

1. Heat a methylbenzoic acid or phenylethanoic acid with sodalime:

$$CH_3C_6H_4COOH(s) + 2NaOH(s) \rightarrow Na_2CO_3(s) + C_6H_5CH_3(l)$$

There are three isomeric methylbenzoic acids:

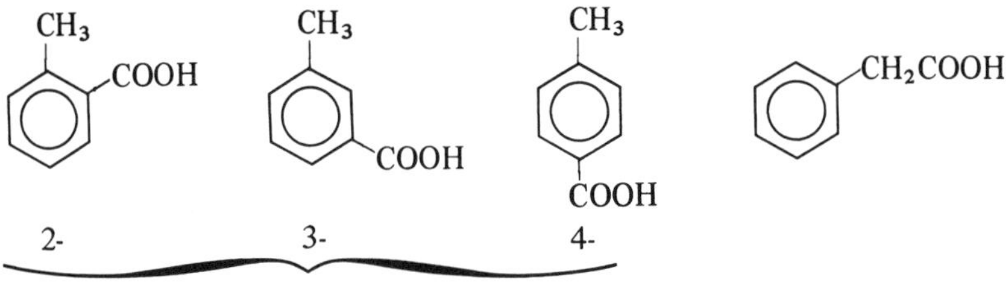

2- 3- 4-

methylbenzoic acids phenylethanoic acid

2. *Friedel-Crafts reaction.* Benzene and a halogenomethane react in the presence of anhydrous aluminium chloride (p. 79):

$$C_6H_6 + ClCH_3 \xrightarrow{AlCl_3} HCl + C_6H_5CH_3$$

Ethylbenzene is formed in this reaction from ethene and benzene; it is used to manufacture phenylethene (p. 55).

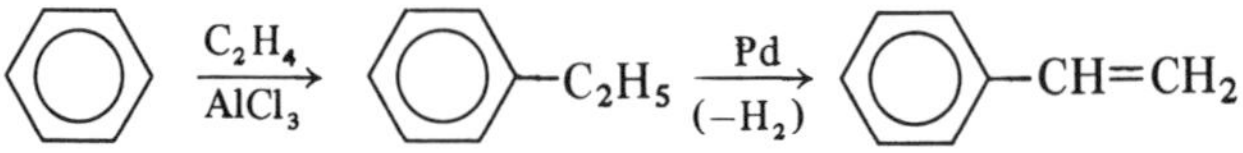

$$C_6H_6 \xrightarrow[AlCl_3]{C_2H_4} C_6H_5{-}C_2H_5 \xrightarrow[(-H_2)]{Pd} C_6H_5{-}CH{=}CH_2$$

3. *Fittig reaction.* Dry halogenoderivatives of benzene and of methane are treated with sodium in dry ether, which is a solvent for the halogen compounds. (Compare the Wurtz reaction, p. 19.)

$$C_6H_5\boxed{Br + 2Na + Br}CH_3 \rightarrow C_6H_5CH_3 + 2NaBr$$

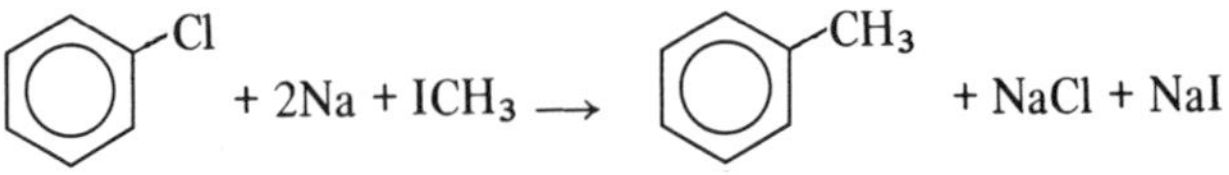

$$C_6H_5Cl + 2Na + ICH_3 \longrightarrow C_6H_5CH_3 + NaCl + NaI$$

Properties of methylbenzene

Physical properties. It is a colourless inflammable liquid with a characteristic smell, and it is poisonous. It boils at 383.3 K, freezes at 178.2 K and its density is 0.87 g cm^{-3} at 298 K. It is practically insoluble in water. It is a good solvent for many organic compounds.

Chemical properties

1. *Combustion.* Methylbenzene burns with a luminous smoky flame (the compound contains 91.3 per cent by mass of carbon):

$$C_6H_5CH_3 + 9O_2 \rightarrow 7CO_2 + 4H_2O$$

2. *Addition reactions of methylbenzene.* Most reactions of methylbenzene involve substitution in the methyl side-chain or the benzene ring, and only one addition reaction of importance occurs.

(a) *Hydrogen.* Methylbenzene vapour and hydrogen combine under pressure and in the presence of nickel at 470 K.

$$C_6H_5CH_3(g) + 3H_2 \rightarrow \underset{\text{methylcyclohexane}}{C_6H_{11}CH_3}$$

3. *Reactions affecting the methyl side-chain.*
(a) *Oxidation.* The side-chain is oxidized to the carboxyl group —COOH by boiling nitric acid, boiling potassium manganate(VII) solution or chromic(VI) acid (sodium dichromate(VI), $Na_2Cr_2O_7$, and concentrated sulphuric acid):

$$C_6H_5{-}CH_3 \xrightarrow[100\,°C]{KMnO_4} \underset{\text{benzoic acid}}{C_6H_5{-}COOH}$$

The carbon ring in methylbenzene is not changed and this is further evidence of its great stability.

Ethylbenzene is also oxidized to benzoic acid:

$$C_6H_5{-}CH_2CH_3 \xrightarrow[H_2SO_4,\ 373\,K]{K_2Cr_2O_7} C_6H_5{-}COOH + 2H_2O + CO_2$$

Heated manganese(IV) oxide and sulphuric acid is a less powerful oxidizing agent and it converts methylbenzene only to an aldehyde:

$$C_6H_5CH_3 + 2[O] \rightarrow H_2O + \underset{\text{benzaldehyde}}{C_6H_5CHO}$$

(b) *Halogenation.* If chlorine is passed into methylbenzene boiling under reflux or into cold methylbenzene in sunlight or ultraviolet light, substitution occurs in the side-chain. The hydrogen atoms of the methyl group are substituted successively, forming hydrogen chloride and the compounds:

$C_6H_5CH_2Cl$	$C_6H_5CHCl_2$	$C_6H_5CCl_3$
(chloromethyl)benzene	(dichloromethyl)benzene	(trichloromethyl)benzene

The chlorination can be stopped at any of the three stages when the calculated increase in mass has occurred.

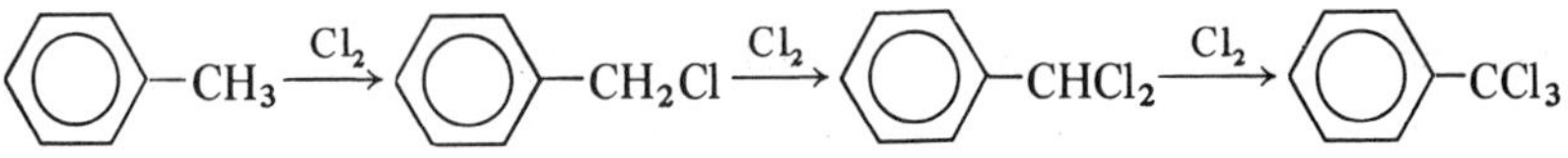

$$C_6H_5\text{-}CH_3 \xrightarrow{Cl_2} C_6H_5\text{-}CH_2Cl \xrightarrow{Cl_2} C_6H_5\text{-}CHCl_2 \xrightarrow{Cl_2} C_6H_5\text{-}CCl_3$$

The free radical mechanism of the substitution is similar to that of the chlorination of methane (p. 31).

The reaction of bromine with methylbenzene is similar to that of chlorine, but iodine has no action.

4. *Substitution reactions affecting the aromatic ring.* There are three possible compounds $C_6H_4XCH_3$ in which the atom or group X is joined to the benzene ring. The carbon atom joined to the methyl group is number 1.

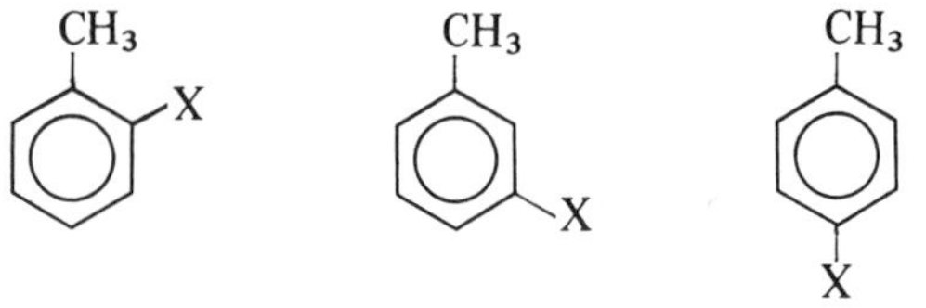

CH₃ / X structures:

2- 3- 4-

Substitution in the benzene ring of methylbenzene produces a mixture of 2- and 4-derivatives, and very little 3-compound. Substitution occurs under milder conditions than those for benzene.

(a) *Halogenation.* Chlorine and bromine substitute hydrogen attached to the ring. A halogen-carrier such as iodine or iron(III) chloride or iron(III) bromide acts as a catalyst. Chlorine forms a mixture of 1-chloro-2-methyl- and 1-chloro-4-methylbenzenes, and further chlorination produces 1-methyl-1,4-dichlorobenzene.

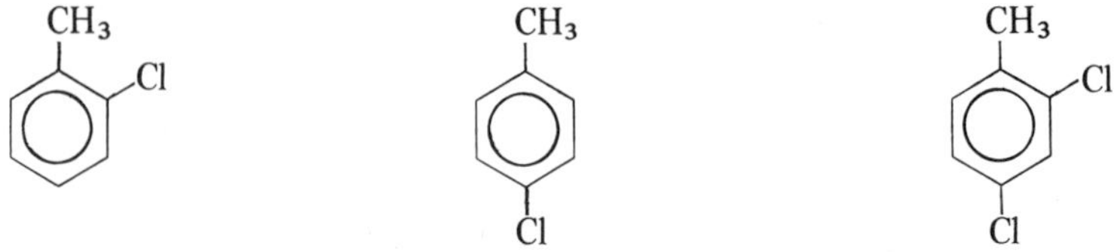

1-chloro-2-methylbenzene 1-chloro-4-methylbenzene 1-methyl-2,4-dichlorobenzene

Bromine forms similar substitution products.

(b) *Sulphuric acid.* Methylbenzene is sulphonated by the cold fuming acid or when heated under reflux with concentrated sulphuric acid at about 470 K. A mixture of 2- and 4-methylbenzenesulphonic acids is formed:

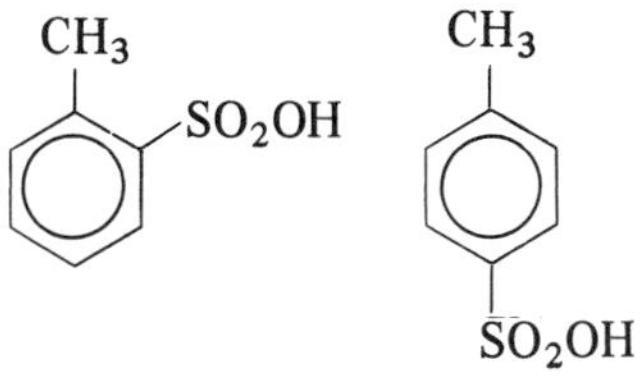

The 2- acid is used in the manufacture of saccharin.

(c) *Nitric acid.* Methylbenzene nitrates more readily than benzene. The nitrating agent can be nitric acid alone or a mixture of concentrated nitric and sulphuric acids. The product is a mixture of 1-methyl-2- and 1-methyl-4-nitrobenzenes.

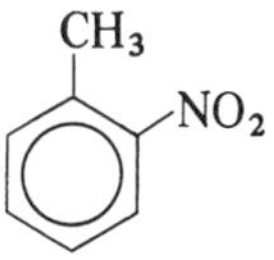

1-methyl-2-nitrobenzene 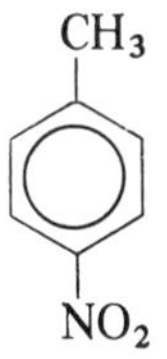 1-methyl-4-nitrobenzene 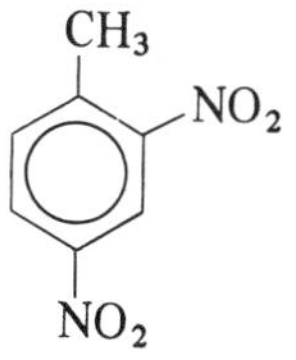 1-methyl-2,4-dinitrobenzene

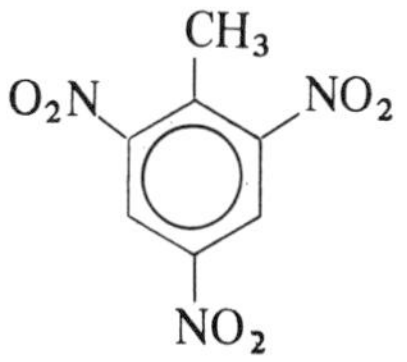

2-methyl-1,3,5-trinitrobenzene (TNT)

Further nitration with a mixture of acids at about 320 K produces 1-methyl-2,4-dinitrobenzene; at 373 K 2-methyl-1,3,5-trinitrobenzene is formed as a yellow solid, which is the common explosive called TNT (trinitrotoluene).

(d) *Halogenoalkanes and acid chlorides* (*Friedel-Crafts reactions*). Methylbenzene is alkylated by halogenoalkanes at room temperatures and by acid chlorides or anhydrides when warm. The catalysts are aluminium chloride or bromide and iron(III) bromide. With bromoethane and ethanoyl chloride the products are:

1-ethyl-4-methylbenzene (4-methylphenyl)ethanone

1-ethyl-4-methylbenzene reacts further to form diethyl- and triethylmethylbenzenes.

Effects on properties of substituents in monosubstituted benzenes

Methylbenzene, phenol and phenylamine ($C_6H_5{-}CH_3$, $C_6H_5{-}OH$ and $C_6H_5{-}NH_2$) are three examples of substituted benzenes which are more reactive than benzene itself. Nitrobenzene, $C_6H_5{-}NO_2$, and benzoic acid, $C_6H_5{-}COOH$, are less reactive than benzene. A benzene molecule has six delocalized electrons moving in π-orbitals around all six carbon atoms. The substituent X in a compound $C_6H_5{-}X$ has an inductive effect (see p. 188) on the six electrons; it either supplies electrons to or withdraws electrons from the ring, i.e. $C_6H_5 \leftarrow X$ or $C_6H_5 \rightarrow X$. A substituent (e.g. $-CH_3$, $-OH$ and $-NH_2$) which supplies electrons to the benzene ring activates the ring and is *ortho*- and *para*-directing. A substituent which withdraws electrons (e.g. $-NO_2$, $-CHO$, $-COOH$, $-CN$ and $-SO_2OH$) deactivates the ring and is *meta*-directing.

The halogen groups Cl— and Br— make the substance less reactive because they are strongly electronegative and withdraw electrons. They are exceptional in that they are *o*- and *p*-directing.

SUMMARY

Production of benzene and methylbenzene

1. Catalytic reforming or platforming of petroleum naphtha.

2. From light oil, a fraction obtained from coal tar produced by destructive distillation of coal.

Benzene ring structure

12 atoms of C_6H_6 are in one plane. The atoms are joined by σ-bonds. 6 delocalized electrons from π-orbitals above and below the ring.

Properties

Reagent/reaction	*Benzene, C_6H_6*	*Methylbenzene, $C_6H_5CH_3$*
Combustion	$CO_2 + H_2O$ in air. Smoky, luminous flames	
Addition		
H_2/Ni, 200 °C	C_6H_{12} cyclohexane	$C_6H_{11}CH_3$ methylcyclohexane
Cl_2, sunlight, boil	$C_6H_6Cl_6$ hexachlorocyclohexane	Substitution occurs
O_3, cold	$C_6H_6(O_3)_3$(s), triozonide	No addition
Substitution in ring		
$Cl_2/AlCl_3$, $FeCl_3$	C_6H_5Cl chlorobenzene	$C_6H_4ClCH_3$, 2- and 4-
$Br_2/AlBr_3$, $FeBr_3$	C_6H_5Br bromobenzene	$C_6H_4BrCH_3$, 2- and 4-
H_2SO_4	Reflux with conc acid Warm with fuming acid $C_6H_5SO_2OH$	Reflux with conc acid Fuming acid even at 0 °C $CH_3C_6H_4SO_2OH$, 2- and 4-
HNO_3/H_2SO_4 (conc)	$C_6H_5NO_2$ at 50 °C $C_6H_4(NO_2)_2$ at 100 °C 1,3-dinitrobenzene	$CH_3C_6H_4NO_2$ at 30 °C $CH_3C_6H_2(NO_2)_3$ at 100 °C 2-methyl-1,3,5-trinitrobenzene
$C_2H_5Br/FeBr_3$	$C_6H_5C_2H_5$ ethylbenzene	$CH_3C_6H_4C_2H_5$, 1,2- and 1,4-
$CH_3COCl/AlCl_3$	$C_6H_5COCH_3$ phenylethanone	$CH_3C_6H_4COCH_3$ (4-methylphenyl)ethanone

Oxidizing agents convert methylbenzene to benzoic acid, C_6H_5COOH. Chlorine (and bromine) in sunlight or with boiling methylbenzene substitute in the side-chain forming $C_6H_5CH_2Cl$, $C_6H_5CHCl_2$ and $C_6H_5CCl_3$. Similar reactions with benzene are not possible.

QUESTIONS

1. Mention briefly three reactions in which benzene reacts as a saturated compound. For each reaction, name the reagent, state the conditions and write a balanced equation.

2. Compare and contrast the reactions of benzene and one named alkane with (a) oxygen, (b) chlorine, and (c) concentrated sulphuric acid.

3. Write full displayed formulae for cyclohexane and cyclohexene. Mention three reactions which show that a molecule of cyclohexene probably contains a double bond.

4. State the conditions under which methylbenzene can be converted in the laboratory to (a) $C_6H_4ClCH_3$, (b) $C_6H_5CH_2Cl$, and (c) C_6H_5COOH. Name these three substances.

5. Discuss the reactions on methylbenzene under various conditions of (a) nitric acid, (b) bromine, and (c) potassium manganate(VII), $KMnO_4$.

6. Compare and contrast the reactions of benzene and either ethene or cyclohexene. How do you account for the similarities and the differences?

7. Outline the evidence which indicates that the carbon atoms of a benzene molecule are arranged in a cyclic structure.

8. An old formula for a benzene molecule shows that it has six C—H bonds, three C—C bonds, and three C=C bonds. The average bond energies are: C—C 346, C=C 610, and C—H 413 kJ mol^{-1}. Use these values to calculate the theoretical enthalpy of formation of benzene. The experimental value is 5500 kJ mol^{-1}. Account for the difference between the theoretical and experimental values.

9. The bond length for C—C bond is 0.154 nm and that for a C=C bond is 0.135 nm. Suggest possible values for the lengths of the carbon–carbon bonds in (a) benzene, (b) cyclohexene, and (c) buta-1,3-diene, $CH_2{=}CH{-}CH{=}CH_2$. Give reasons for your answers.

10. Some aliphatic hydrocarbons are said to be ***unsaturated*** and others ***saturated.*** Explain the meanings of these two words, and refer to both structures and reactions of two unsaturated and two saturated aliphatic hydrocarbons. In addition to the four examples you choose, refer to the reactions of cyclohexane and hex-1-ene, which are isomers of molecular formula C_6H_{12}. Indicate the possible mechanisms of one substitution and one addition reaction you describe.

Chapter 6
HALOGEN DERIVATIVES OF ALKANES

MONOHALOGENOALKANES

The monohalogen derivatives of alkanes have the general formula $C_nH_{2n+1}Hal$ in which Hal represents a halogen atom (F, Cl, Br or I). The functional group is C—Hal (C—F, C—Cl, C—Br or C—I). The compounds contain alkyl groups, C_nH_{2n+1}—, and are sometimes called *alkyl halides.* The carbon-fluorine bond is so strong that fluoroalkanes are very unreactive. However, the chloro-, bromo- and iodo-derivatives are reactive and are used in many organic reactions.

One only halogenomethane, CH_3Hal, and one halogenoethane, C_2H_5Hal, can exist for each halogen. For example, the formula of iodomethane is:

```
                    H
                    |
   CH3I    or    H—C—I
                    |
                    H
```

and that of bromoethane is:

```
                                     H   H
                                     |   |
C2H5Br    or    CH3CH2Br    or    H—C——C—Br
                                     |   |
                                     H   H
```

Two isomers exist of each halogenopropane, C_3H_7Hal:

$CH_3CH_2CH_2Br$ (1-bromopropane) and $CH_3CHBrCH_3$ (2-bromopropane)

Four isomers exist of each compound C_4H_9Hal:

C_4H_9I or $CH_3CH_2CH_2CH_2I$

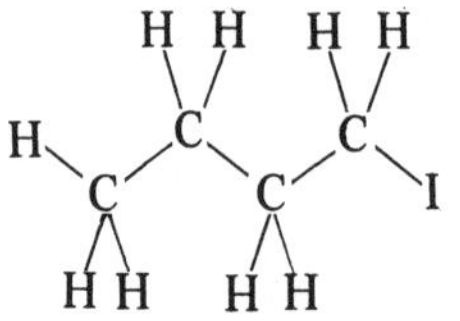

1-iodobutane

$C_2H_5CHICH_3$ or $CH_3CH_2CHICH_3$

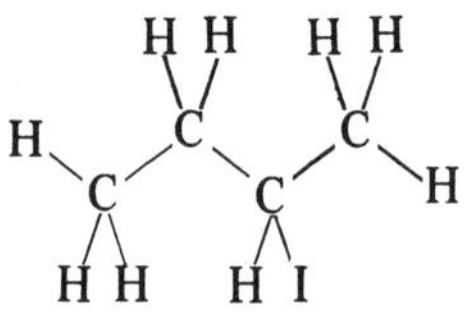

2-iodobutane

$(CH_3)_2CHCH_2I$

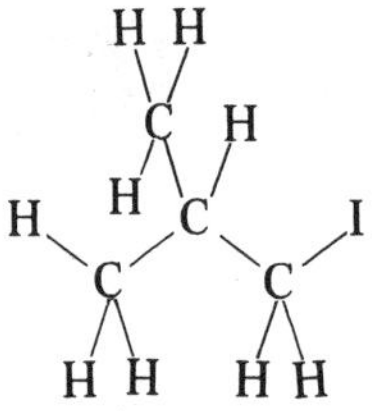

1-iodo-2-methylpropane

$(CH_3)_3CI$

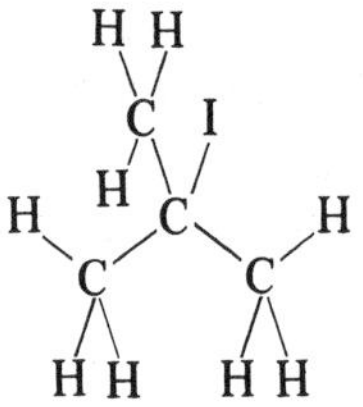

2-iodo-2-methylpropane

The two isomers $CH_3CH_2CH_2CH_2I$ and $(CH_3)_2CHCH_2I$ are primary halogenoalkanes, $C_2H_5CHICH_3$ is a secondary compound, and $(CH_3)_3CI$ is a tertiary compound. The general formulae of the three types of halogenoalkanes are:

$R{-}CH_2{-}Hal$ (primary)

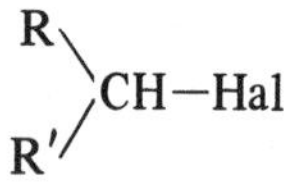

$RR'CH{-}Hal$ (secondary)

$RR'R''C{-}Hal$ (tertiary)

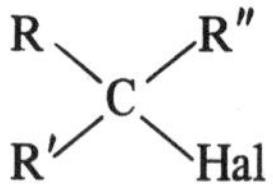

Hal = Cl, Br or I, and R, R′ and R″ = alkyl group.

Formation or preparation of monohalogenoalkanes

1. *Halogenation of alkanes.* Chlorine and bromine, but not iodine, replace the hydrogen of alkanes in the presence of diffused light (p. 26):

$$CH_4(g) + Cl_2(g) \rightarrow HCl(g) + CH_3Cl(g)$$

The method has no value in the laboratory because it yields mixtures of products, e.g. CH_2Cl_2, $CHCl_3$ and CCl_4. Chlorination of methane at 670 K is used to manufacture chloromethane.

2. *From an alcohol and halogen acid* (replacement of —OH by —Hal). An alcohol and halogen acid react reversibly when heated:

$$C_2H_5OH(l) + HBr(aq) \rightleftharpoons H_2O + C_2H_5Br(l)$$

A dehydrating agent is necessary to remove the water and allow the reaction to proceed as far as possible to the right. The hydrogen bromide is prepared, in the presence of the ethanol, by reaction between concentrated sulphuric acid and solid sodium bromide or potassium bromide. Hydrogen bromide solution or hydrobromic acid cannot be used because it contains water. Excess sulphuric acid is used and it acts as the dehydrating agent.

$$H_2SO_4(l) + KBr(s) \rightarrow HBr + KHSO_4$$

3. *From an alcohol and phosphorus halide or sulphur dichloride oxide*, SCl_2O (replacement of —OH by —Hal). The trichloride and pentachloride of phosphorus react readily with alcohols at ordinary temperatures:

$$3C_2H_5OH(l) + PCl_3(l) \rightarrow 3C_2H_5Cl + \underset{\text{phosphonic acid}}{P(OH)_3 \text{ or } H_2PHO_3}$$

$$C_2H_5OH(l) + PCl_5(s) \rightarrow C_2H_5Cl + HCl + \underset{\text{phosphorus trichloride oxide}}{PCl_3O}$$

Bromoalkanes and iodoalkanes are usually prepared from the alcohol, red phosphorus and bromine or iodine respectively. The phosphorus and halogen combine to form phosphorus halide, which reacts with the alcohol:

$$2P + 3Br_2 \rightarrow 2PBr_3$$

$$\begin{array}{l} C_2H_5OH \qquad\quad Br \\ \qquad\qquad\qquad\;\; / \\ C_2H_5OH + P{-}Br \rightarrow 3C_2H_5Br + \underset{\text{phosphonic acid}}{P(OH)_3 \text{ or } H_2PHO_3} \\ \qquad\qquad\qquad\;\; \backslash \\ C_2H_5OH \qquad\quad Br \end{array}$$

Alcohols and sulphur dichloride oxide form a chloroalkane and gaseous byproducts:

$$C_3H_7OH(l) + SCl_2O(l) \rightarrow C_3H_7Cl(l) + SO_2(g) + HCl(g)$$

4. *Addition of a halogen acid to an alkene.* Hydriodic acid and hydrobromic acid add readily at ordinary temperatures to alkenes, but hydrochloric acid adds very slowly (p. 47).

$$H_2C{=}CH_2(g) + HBr(aq) \rightarrow CH_3CH_2Br(l)$$

The method has no practical value for simple halogenoalkanes.

To prepare bromoethane, C_2H_5Br

Principle. Concentrated sulphuric acid is added to ethanol and the mixture is cooled. Sodium or potassium bromide is added (method 2 above), allowed to react, and the mixture is distilled. The impure bromoethane is purified and re-distilled.

Mixing the reagents. Add ethanol or industrial spirit (15 cm^3) and water (10 cm^3) to a beaker. Add concentrated sulphuric acid (16 cm^3) from a measuring cylinder a little at a time to the alcohol. Cool the beaker in a cold water bath because much heat is evolved during the mixing. Shake the mixture gently from time to time to ensure that the dense acid mixes with the alcohol.

Transfer the mixture to a distillation flask of capacity at least 100 cm^3 and preferably larger. Use a pestle and mortar to grind sodium bromide (14 g) or potassium bromide (16 g) and add the powder to the cold mixture. A little bromine is liberated and a slight reddish colour appears.

Distilling. Support the distillation flask on a tripod and gauze and arrange the rest of the apparatus as in Fig. 6.1. Warm the mixture with a low flame so that the product distils over slowly and steadily. The mixture tends to froth; stop heating for a while if froth seems about to pass into the condenser. Continue distillation until no more colourless oily drops of bromoethane pass into the receiver and collect as a lower layer.

Purifying the bromoethane. Impurities present include water, ethanol, hydrogen bromide, bromine, and possibly ethoxyethane (formed from ethanol and sulphuric acid).

(a) Pour the whole distillate into a separating funnel and decant the upper aqueous layer which is not required. Retain the lower layer of bromoethane.

(b) To the impure bromoethane add an equal volume of sodium carbonate solution, about 2 M. Close the top of the funnel with a stopper, invert the funnel, shake gently, and release any carbon dioxide formed by opening the tap. Repeat the procedure until no more gas is evolved. The carbonate reacts with any hydrogen bromide and bromine present in the bromoethane. Again decant the upper aqueous layer.

(c) Shake the bromoethane with an equal volume of water to remove any sodium carbonate. Run the moist bromoethane into a small dry conical flask.

(d) Add a few large pieces of calcium chloride to the bromoethane and stopper the flask. Allow the mixture to stand for a day at least. The calcium chloride removes water. Decant the clear bromoethane into a small distillation flask.

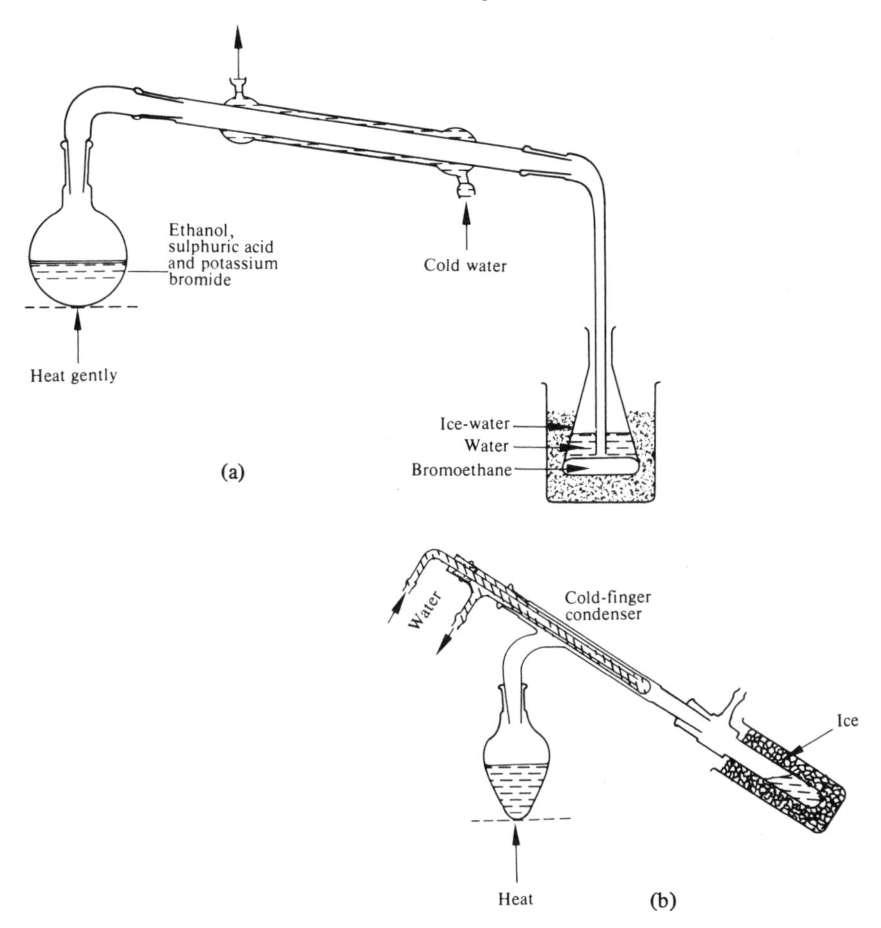

Fig. 6.1. Preparation of bromoethane

(e) Distil the bromoethane. Collect the fraction boiling between 310 and 312 K (but not the first and last drops). Keep the specimen in a sealed tube and store in a dark place because light causes decomposition with liberation of bromine. The yield is about 10 g.

Electronegativities of elements

The electronegativity of an element is a measure of the electron-attracting power of its atom when in a molecule. A fluorine atom, electronic configuration 2,7, attracts an electron most readily and forms a fluoride ion F^- with a noble gas configuration, 2,8. All the elements are placed on a numerical electronegativity scale, devised by Pauling, which varies from 4.0 for fluorine to 0.7 for caesium.

Non-metals have high numbers (usually more than 2.0) and metals have low numbers, usually 1.5 or less. Some values are:

Fluorine	4.0	Chlorine	3.0	Bromine	2.8	Iodine	2.4
Lithium	1.0	Sodium	0.9	Potassium	0.8		
Hydrogen	2.1	Carbon	2.5	Nitrogen	3.0	Oxygen	3.5

The electronegativities of carbon and hydrogen are similar. In a carbon-hydrogen bond, C—H, both atoms attract the shared pair equally, and the covalent bond is non-polar. Halogens, except iodine, are more electronegative than carbon. In a carbon-halogen bond there is a movement of negative charge towards the halogen atom. C—Cl and C—Br bonds are polar and have dipole moments $C^{\delta+}—Hal^{\delta-}$.

In halogenoalkane molecules, the halogen atom acquires more than an equal share of the electrons of the covalent bond, and the polar molecules can be represented:

$$CH_3—\overset{H}{\underset{H}{C^{\delta+}}}—Cl^{\delta-} \text{ or } CH_3—\overset{H}{\underset{H}{C}}\rightarrow Cl \quad H—\overset{H}{\underset{H}{C}}—\overset{H}{\underset{H}{C^{\delta+}}}—Br^{\delta-} \text{ or } H—\overset{H}{\underset{H}{C}}—\overset{H}{\underset{H}{C}}\rightarrow Br$$

The small negative charge on the halogen atom and the small positive charge on a carbon atom account for many properties and reactions of halogenoalkanes.

For example, the strongly polar chloroalkane molecules attract each other and energy is needed to overcome this attraction. The boiling points of chloroalkanes are therefore greater than those of alkanes.

Substitution (nucleophilic) reactions of halogenoalkanes

In a halogenoalkane molecule, the carbon atom with a small positive charge reacts readily with nucleophiles such as OH^-, CN^- and NH_3. The two ions have negative charges and the ammonia molecule has a lone pair of electrons, and therefore these nucleophiles are attracted to the positive carbon atom. The following equations, in which *a curved arrow represents a movement of a pair of electrons*, represent the substitution reactions:

$$HO^- + H—\overset{H}{\underset{H}{C}}—Br \xrightarrow[\text{step}]{\text{rate-determining}} \left[H—O\cdots\overset{H\ \ H}{\underset{H}{C}}\cdots Br \right]^- \xrightarrow{\text{fast}} H—O—\overset{H}{\underset{H}{C}}—H + Br^-$$

transition state

i.e. $$HO^- + CH_3Br \xrightarrow{\text{slow}} [HO\cdots CH_3\cdots Br]^- \xrightarrow{\text{fast}} HO{-}CH_3 + Br^-$$

$$HO^- + \overset{\delta+}{CH_3}{-}\overset{\delta-}{Br} \rightarrow HO{-}CH_3 + Br^-$$

The rate of reaction is determined by collision between two particles, a hydroxide ion and a halogenoalkane molecule. This process is called an S_N2 reaction: S for substitution, N for nucleophilic (the OH^- is a nucleophilic reagent), and 2 for two particles. The *transition state* particle, or *activated complex*, exists only temporarily.

$$NH_3 + H{-}\underset{CH_3}{\overset{H}{C}}{-}Hal \rightarrow \left[H_3N\!:\cdots\underset{CH_3}{\overset{H\ \ H}{C}}\cdots Hal \right] \rightarrow CH_3CH_2NH_2 + Hal^- + H^+$$

The nucleophile approaches the positively charged carbon atom and supplies electrons, and thus enables the halogen atom to leave with an electron pair, i.e. as a halide ion. The general equation, in which Nu^- represents the nucleophile is:

$$Nu^- + R{-}Hal \rightarrow Nu{-}R + Hal^-$$

Tests on monohalogenoalkanes

Use either iodomethane, bromoethane or iodoethane in the following tests because they are liquid at room temperatures.

1. *Silver nitrate.* Add a few drops of alcoholic silver nitrate solution to the freshly prepared halogenoalkane. Observe if a precipitate forms. Add a few drops of dilute nitric acid and warm gently. Observe if any change occurs. (Iodomethane and silver nitrate form a white precipitate that is a complex compound of the two and is not silver iodide, which is yellow.)

2. *Sodium hydroxide.* Warm the halogenoalkane with about four times its volume of aqueous sodium hydroxide (about 4 M) under reflux. Observe if the oily drops gradually disappear.

Acidify a small portion of the product with dilute nitric acid and then add aqueous silver nitrate. Observe the precipitate, if any.

3. *Potassium hydroxide (alcoholic).* Add about 5 cm^3 of alcoholic potassium hydroxide to about 10 drops of bromoethane or iodoethane in a test-tube. Fit with a stopper and delivery tube and arrange to collect any gas over water. Warm the mixture gently. Ignite a tube of any gas formed.

4. *Sodium.* Place clean pieces of sodium in a small dry flask. Add the halogenoalkane, drop by drop, from a dropping funnel to the sodium and arrange to collect any gas over water. Ignite a tube of the gas formed.

Properties of halogenoalkanes

Physical properties. Chloromethane, bromomethane and chloroethane are gases at ordinary temperatures. The other compounds are colourless liquids with a sweet sickly smell similar to that of chloroform.

Halogenoalkane	*Formula*	*B.p./K*	*Density/g cm^{-3}*
Chloromethane	CH_3Cl	249	0.92 at 273 K
Bromomethane	CH_3Br	277	1.68 at 273 K
Iodomethane	CH_3I	316	2.28 at 298 K
Chloroethane	C_2H_5Cl	285.4	0.90 at 273 K
Bromoethane	C_2H_5Br	311.5	1.46 at 298 K
Iodoethane	C_2H_5I	345.5	1.94 at 298 K

The halogenoalkanes are almost insoluble in water but are soluble in ethanol, ether and some other organic solvents.

Chemical properties

1. *Combustion.* The compounds burn with a greenish flame.

2. *Substitution reactions.*

(a) *Substitution of* —Hal *by* —OH *to form an alcohol.* Halogenoalkanes react with *aqueous* sodium hydroxide, potassium hydroxide or 'moist silver oxide', which probably reacts as silver hydroxide, AgOH:

$$C_2H_5{-}Br + H{-}OH \rightarrow C_2H_5{-}OH + HBr \text{ (removed by alkali)}$$

The C—Hal bond appears to break and the reaction is usually called a hydrolysis. However the active reagent is OH^- and the true reaction is:

$$C_2H_5Br(l) + OH^-(aq) \rightarrow C_2H_5OH(aq) + Br^-(aq) \text{ (removed by alkali)}$$

The reaction has no application with simple halogenoalkanes, which are prepared from alcohols in the first place, but is applied to complex compounds.

(b) *Substitution of* —Hal *by* —CN *to form nitrile.* Halogenoalkanes react with warm potassium cyanide dissolved in a mixture of ethanol and water. The ethanol is a solvent for the halogen compound and the water for the cyanide.

$$C_2H_5I + CN^- \rightarrow I^- + \underset{\text{propanenitrile}}{C_2H_5CN}$$

$$R{-}Hal + CN^- \rightarrow Hal^- + \underset{\text{a nitrile}}{R{-}CN}$$

R represents an alkyl group. Since the nitrile can be hydrolysed by acid or alkali to a carboxylic acid, this reaction enables an alcohol to be converted to a halogenoalkane and then to an acid. The acid molecule contains one more carbon atom than the alcohol molecule, e.g.

$$R{-}OH \rightarrow R{-}Hal \xrightarrow{KCN} R{-}CN \xrightarrow{H_2O} R{-}COOH$$

$$CH_3OH \rightarrow CH_3I \longrightarrow CH_3CN \longrightarrow CH_3COOH$$

(c) *Substitution of* —Hal *by amino group* —NH_2 *to form amines.* Halogeno-alkanes at about 373 K under pressure (e.g. in a sealed tube) react with excess aqueous or alcoholic ammonia solution. The first reaction is:

$$R{-}Hal + H{-}NH_2 \rightarrow H{-}Hal + \underset{\text{an amine}}{R{-}NH_2}$$

$$C_2H_5Br + NH_3 \rightarrow HBr + \underset{\text{ethylamine}}{C_2H_5NH_2}$$

The hydrogen halide formed reacts with excess ammonia. Further reactions can occur, and the second and third hydrogen atoms of the ammonia molecule are replaced, forming amines R_2NH and R_3N. The reaction is the *alkylation of ammonia*; the halogenoalkane acts as an *alkylating agent* (p. 80).

(d) *Substitution of* —Hal by —OR (R = *alkyl group*). This is *Williamson's ether synthesis.* Halogenoalkanes react with sodium ethoxide (formed from sodium and ethanol) or other sodium alkoxide:

$$R{-}Hal + Na{-}OC_2H_5 \rightarrow NaHal + \underset{\text{an ether}}{R{-}OC_2H_5}$$

$$C_2H_5Br + NaOC_2H_5 \rightarrow NaBr + \underset{\text{ethoxyethane}}{C_2H_5OC_2H_5}$$

This reaction proves that ethoxyethane contains two ethyl groups attached to an oxygen atom. The two alkyl groups need not be the same:

$$CH_3I + NaOC_3H_7 \rightarrow NaI + \underset{\text{methoxypropane}}{CH_3OC_3H_7}$$

In all substitution reactions iodoalkanes are more reactive than bromoalkanes, which are more reactive than chloroalkanes. If R is the same alkyl group, the order of reactivity is:

$$RI > RBr > RCl.$$

3. *Elimination of elements of hydrogen halides (dehydrohalogenation).* Halogenoalkanes react with *alcoholic* potassium hydroxide or sodium hydroxide:

$$R{-}CH_2{-}CH_2{-}I + KOH(alc) \rightarrow KI + H_2O + R{-}CH{=}CH_2$$

R = H or alkyl in this equation. Other equations are on p. 45. Clearly the reaction cannot be used with halogenomethanes because there must be at least two carbon atoms in the molecule. Halogenoethanes yield only about 2 per cent of ethene.

4. *Reduction to alkanes.* A dry aluminium-mercury couple or zinc-copper couple reacts with dry methanol or ethanol (weak acids (p. 133)) which reduces dry halogenoalkanes to alkanes at ordinary temperatures (p. 19):

$$C_2H_5Br + H^+(alc) + Zn(s) \rightarrow Br^- + Zn^{2+} + C_2H_6(g)$$

5. *Conversion to higher alkanes (Wurtz reaction).* Sodium reacts at ordinary temperatures with dry halogenoalkanes dissolved in dry ether:

$$C_2H_5Br + 2Na + BrC_2H_5 \rightarrow 2NaBr + C_2H_5C_2H_5 \quad \text{or} \quad \underset{\text{butane}}{C_4H_{10}}$$

$$C_2H_5Br + 2Na + ICH_3 \rightarrow NaBr + NaI + C_2H_5CH_3 \quad \text{or} \quad \underset{\text{propane}}{C_3H_8}$$

DIHALOGENOALKANES

Dihalogen derivatives of the alkanes have the general formula $C_nH_{2n}X_2$ or $C_nH_{2n}XY$ in which X and Y represent halogen atoms. Two isomers exist of each dihalogenoethane, $C_nH_{2n}X_2$, e.g.

CH_2BrCH_2Br or

```
    H   H
    |   |
H — C — C — H
    |   |
    Br  Br
```

1,2-dibromoethane

CH_3CHBr_2 or

```
    H   H
    |   |
H — C — C — Br
    |   |
    H   Br
```

1,1-dibromoethane

In the symmetrical compound the halogen atoms are attached to different carbon atoms; in the unsymmetrical compound they are attached to the same carbon atom.

1,1-dichloroethane is obtained by the action of phosphorus pentachloride on ethanal:

$$CH_3CHO(l) + PCl_5(s) \rightarrow CH_3CHCl_2(l) + PCl_3O(l)$$

1,1-dihalogenoethanes are also obtained by addition of hydrogen halides to ethyne (p. 64):

$$HC{\equiv}CH(g) + 2HBr(g \text{ or } aq) \rightarrow CH_3CHBr_2(l)$$

1,2-dibromoethane, CH_2BrCH_2Br, is prepared by bubbling ethene through liquid bromine:

$$\begin{matrix} CH_2 & & Br & & CH_2Br \\ \| & + & | & \rightarrow & | \\ CH_2 & & Br & & CH_2Br \end{matrix}$$

Properties of 1,2-dibromoethane

It is a colourless liquid with a pleasant smell. It boils at 405 K and its density is 2.2 g cm^{-3}. It is almost insoluble in water.

The reactions of dibromoethane with alkalis and potassium cyanide resemble those of bromoethane. Boiling *aqueous* alkalis hydrolyse it to ethane-1,2-diol:

$$\begin{matrix} CH_2Br & & HOH & & CH_2OH & \\ | & + & & \rightarrow & | & + 2HBr \text{ (reacts with alkali)} \\ CH_2Br & & HOH & & CH_2OH & \end{matrix}$$

(The isomer CH_3CHBr_2 hydrolyses to ethanal, CH_3CHO; possibly the unstable compound $CH_3CH(OH)_2$ is first formed.) *Alcoholic* potassium hydroxide eliminates the elements of hydrogen bromide:

$$\begin{matrix} CH_2Br & & CH_2 & & CH \\ | & \xrightarrow[\text{alcoholic}]{KOH} & \| & \xrightarrow[(-HBr)]{KOH} & ||| \\ CH_2Br & & CHBr & & CH \end{matrix}$$

Bromoethene is the first product and then ethyne.

Alcoholic potassium cyanide replaces both halogen atoms:

$$CH_2BrCH_2Br + 2KCN(alc) \rightarrow 2KBr + CH_2CNCH_2CN$$

Uses of halogenoalkanes

Chloroethane is used to manufacture tetraethyllead $Pb(C_2H_5)_4$, which is added to petrol to reduce 'knocking' in engines (p. 36). Chloroethane is also used as a local anaesthetic. It is a liquid when under pressure; the liquid is sprayed on the skin, where it evaporates quickly and cools the nerve endings so much that pain cannot be felt.

Triiodomethane is used as an antiseptic. Unlike iodine, it does not irritate the skin. Trichloromethane is used as a solvent.

Poly(tetrafluoroethene), PTFE, is a polymer of formula $(C_2F_4)_n$. It is a white unreactive solid and is an excellent electrical insulator. It has the lowest coefficient of friction known. It is used to coat non-stick cooking vessels, to prevent food sticking to them during heating, and to coat rollers of printing presses. Because it is inert and does not soften even at 670 K, it is used to join pipes used in chemical laboratories and plants.

Dichlorodifluoromethane, CCl_2F_2, is the refrigerant Freon. It is easily liquefied by pressure. It has no smell and is not corrosive, unlike liquid ammonia and liquid sulphur dioxide which were once used in refrigerators. It is in common use as an aerosol propellant in insecticides.

SUMMARY

Monohalogenoalkanes

$C_nH_{2n+1}X$ (X = F, Cl, Br or I). Functional group: C—X, strongly polarized $\overset{\delta+}{C}—\overset{\delta-}{X}$ or C→X.

Preparation

Alcohol + halogen acid (e.g. KBr + conc. H_2SO_4), or PBr_3 (i.e. P + Br_2) or SCl_2O.

Properties

(of C_2H_5Br, bromoethane)

Combustion	$CO_2 + H_2O$ + halogen acid in air
Substitutions	
NaOH(aq), KOH(aq) or moist Ag_2O	Hydrolysis to C_2H_5OH
KCN (ethanol-water), warm	C_2H_5CN propanenitrile
NH_3(aq or alc), heat, pressure	$C_2H_5NH_2$ ethylamine
C_2H_5ONa	$C_2H_5OC_2H_5$ (Williamson reaction) ethoxyethane (ether)
KOH(alc), reflux	$CH_2{=}CH_2$ (dehydrohalogenation) ethene
Al/Hg or Zn/Cu (ethanol)	C_2H_6 (reduction) ethane
Na (dry ether)	$C_2H_5C_2H_5$ (Wurtz reaction) butane

Dihalogenoalkanes

CH_2BrCH_2Br, 1,2-dibromoethane, reacts as above with OH^-(aq) and KCN(alc) forming CH_2OHCH_2OH, ethane-1,2-diol, and CH_2CNCH_2CN.

QUESTIONS

1. Describe, with full experimental details, how ethanol may be converted to bromoethane.

2. How, and under what conditions, does bromoethane react with (a) sodium hydroxide, (b) sodium, (c) potassium cyanide, and (d) sodium ethoxide?

3. Outline the preparation of either trichloroethane or triiodoethane. What is the action of potassium hydroxide solution on the substance chosen?

4. Name and write the displayed formulae for two dichloroethanes. Describe a chemical test by which you would distinguish between the two.

5. Describe the preparation from ethanol of 1,2-dibromoethane. What is the action on the dibromo-compound of (a) sodium hydroxide, and (b) silver nitrate, under various conditions?

6. An unsaturated hydrocarbon, $C_{10}H_{16}$, reacts with bromine to form a saturated product and no hydrogen bromide. The product contains 70.2 per cent of bromine by mass. Show that the hydrocarbon probably contains two double bonds per molecule. (C = 12, H = 1, Br = 80.)

7. An iodoalkane *X* contains, by mass, 26.1 per cent of carbon, 4.9 per cent of hydrogen, and the rest is iodine. *X* reacts with alcoholic potassium hydroxide to form a hydrocarbon *Y*, containing 85.7 per cent of carbon. Oxidation of *Y* yielded carbon dioxide, water, and propanone (CH_3COCH_3). Hydriodic acid added on to *Y* to form an isomer of *X*. Suggest structures for *X, Y* and *Z*. (I = 127.)

8. In the preparation of 1-bromobutane, the starting materials are 10 g of sodium bromide, 10 cm^3 of concentrated sulphuric acid, 10 cm^3 of water, and 7.5 cm^3 of butanol. (a) In what order and with what precautions should these materials be introduced into a 50 cm^3 flask. (b) After thorough mixing, the contents of the flask are boiled. Draw a sketch of the arrangement you would use. (c) After the boiling is completed, the mixture in the flask is allowed to cool and is then distilled, the distillate being collected in a measuring cylinder. (i) Why is the mixture allowed to cool before being distilled? (ii) How is it possible to tell when the distillation is complete? (d) The distillate consists of

two liquid layers. What is the chief impurity in the organic layer and how can this be removed? (e) What further purification processes are necessary to ensure that the final product is pure? (L.)

9. An alcohol of formula $CH_3\underset{\displaystyle OH}{\underset{|}{C}}HCH_2CH_3$ is passed in the vapour state over heated broken porcelain. Each molecule of the alcohol loses one molecule of water. (a) Name the alcohol. (b) Give the formula and name two possible products of the above reaction. (c) Give the formula and name the products formed when the two compounds in (b) combine with (i) bromine, and (ii) hydrobromic acid.

10. (a) Hot concentrated alcoholic potassium hydroxide can eliminate hydrogen bromide from the compound $CH_3CH_2CHBrCH_3$. Suggest and name two possible products. (b) Write the structural formulae and names of all possible products formed by elimination of hydrogen bromide from the dibromoalkane $CH_3CHBrCHBrCH_3$.

11. Suggest reasons for the following: (a) Trichloromethane will not burn continuously in air but 1-chlorododecane, $CH_3(CH_2)_{10}CH_2Cl$, will do so.
(b) Poly(1,1-dichloroethene) is crystalline but poly(chloroethene) is amorphous.
(c) The hydrolysis of chloroethane is a nucleophilic reaction but the addition of halogens to ethene is an electrophilic reaction.

Chapter 7
AROMATIC HALOGEN COMPOUNDS

HALOGENOBENZENES

The monohalogen derivatives of benzene have the general formula C_6H_5Hal. The functional group is C—Hal but its properties are modified by the influence on it of the benzene ring.

C_6H_5Cl or [benzene ring]—Cl
chlorobenzene

Preparation of monohalogenobenzenes

Benzene reacts with chlorine and bromine with a catalyst (a *halogen carrier*) such as iron(III) halide, iodine, and aluminium chloride and in the absence of sunlight. Iodination of benzene is not carried out because the yield is low, and fluorination is impossible because fluorine reacts too vigorously.

Chlorobenzene. Pass chlorine through dry benzene in the presence of iron(III) chloride until the calculated increase in mass has occurred:

$$C_6H_6(l) + Cl_2(g) \xrightarrow{FeCl_3} HCl(g) + C_6H_5Cl(l)$$

$$\text{[benzene ring]} \xrightarrow[FeCl_3\ (-HCl)]{Cl_2} \text{[benzene ring]—Cl}$$

Separate the chlorobenzene, b.p. 405 K, by fractional distillation from any benzene left unchanged.

Bromobenzene. Mix bromine, benzene and a few iron filings in a flask with a reflux condenser and warm to about 320 K. When reaction seems to be finished, boil the mixture gently to drive off any bromine. Bromobenzene boils at 429 K.

$$C_6H_6 \xrightarrow[FeBr_3 \; (-HBr)]{Br_2} C_6H_5\text{—}Br$$

Chlorobenzene is *not* prepared from phenol, C_6H_5OH, and PCl_5.

Properties of halogenobenzenes

Physical properties. They are colourless liquids, denser than and almost insoluble in water.

Chemical properties. The halogen atom is firmly attached to the benzene ring and the compounds are unreactive. The halogen atom cannot be replaced by OH, CN or NH_2 by reaction with alkalis, potassium cyanide and ammonia solution, and therefore it differs markedly from the reactive halogen atom in halogenoalkanes. A molecule can be represented by:

$$C_6H_5^{\delta+}\text{—}Cl^{\delta-}$$

The halogen deactivates the ring but is *ortho-/para*-directing (p. 86).

1. *Reduction to benzene.* Sodium amalgam or lithium tetrahydridoaluminate(III), $Li[AlH_4]$, reduces to benzene:

$$C_6H_5Cl + [H] \text{ (from reducing agent)} \rightarrow HCl + C_6H_6$$

2. *Fittig reaction.* A halogenobenzene and a halogenoalkane react with sodium in dry ether as solvent (p. 82).

$$C_6H_5Br + 2Na + BrC_2H_5 \rightarrow 2NaBr + \underset{\text{ethylbenzene}}{C_6H_5C_2H_5}$$

The two reactions above resemble those of halogenoalkanes.

3. *Nitration.* A mixture of concentrated nitric and sulphuric acids converts them to a mixture of 2- and 4-chloronitrobenzenes:

$$C_6H_5Cl + HNO_3 \rightarrow O_2NC_6H_4Cl + H_2O$$

4. *Sulphonation.* Concentrated sulphuric acid forms a mixture of 2- and 4-halogenobenzenesulphonic acids:

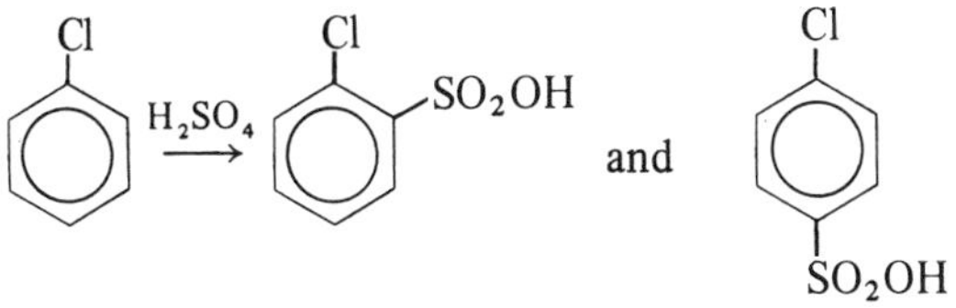

$$C_6H_5Cl \xrightarrow{H_2SO_4} \text{2-}ClC_6H_4SO_2OH \text{ and } \text{4-}ClC_6H_4SO_2OH$$

5. *Halogenation.* Chlorine, in the presence of a halogen carrier, reacts with chlorobenzene to form 1, 2- and 1,4-dichlorobenzenes and then 1,2,4-trichlorobenzene:

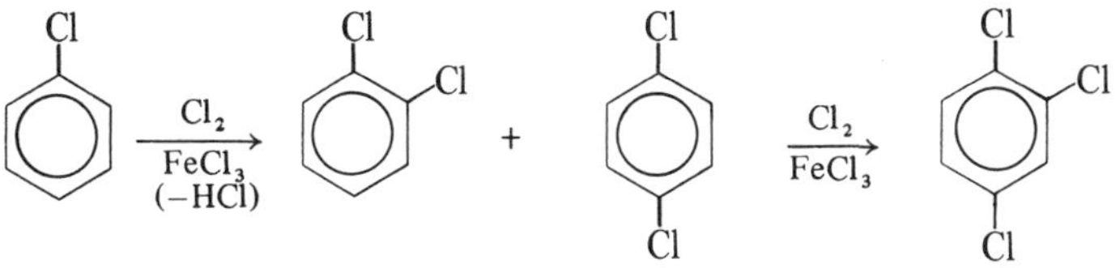

$$C_6H_5Cl \xrightarrow[\text{FeCl}_3\ (-\text{HCl})]{\text{Cl}_2} \text{1,2-}C_6H_4Cl_2 + \text{1,4-}C_6H_4Cl_2 \xrightarrow[\text{FeCl}_3]{\text{Cl}_2} \text{1,2,4-}C_6H_3Cl_3$$

Bromine reacts in a similar way.

HALOGENOMETHYLBENZENES

Each halogen can form four isomeric monohalogen derivatives of methylbenzene, C_7H_7Hal; in one of these the halogen atom is attached to the carbon atom of the side-chain, $C_6H_5CH_2Hal$, and the compounds contain the group $C_6H_5CH_2—$, e.g.

$$C_6H_5CH_2Cl$$
(chloromethyl)benzene

In the other isomers the halogen atom is attached to the benzene ring, $C_6H_4HalCH_3$:

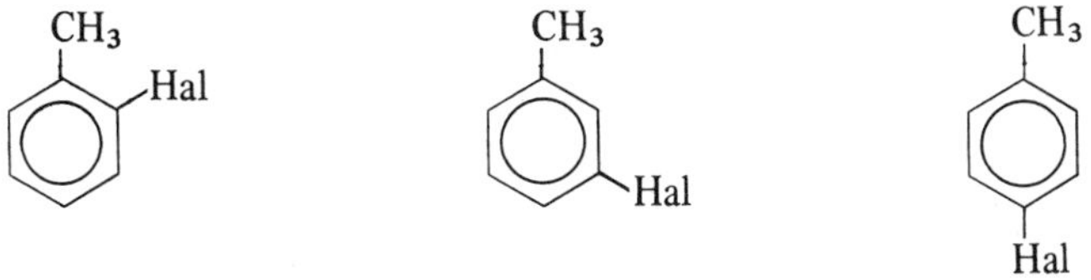

$$\text{2-}CH_3C_6H_4Hal \qquad \text{3-}CH_3C_6H_4Hal \qquad \text{4-}CH_3C_6H_4Hal$$

The hydrogen atoms of the side chain can be substituted successively, e.g.

$C_6H_5CH_2Br$ $C_6H_5CHBr_2$ $C_6H_5CBr_3$

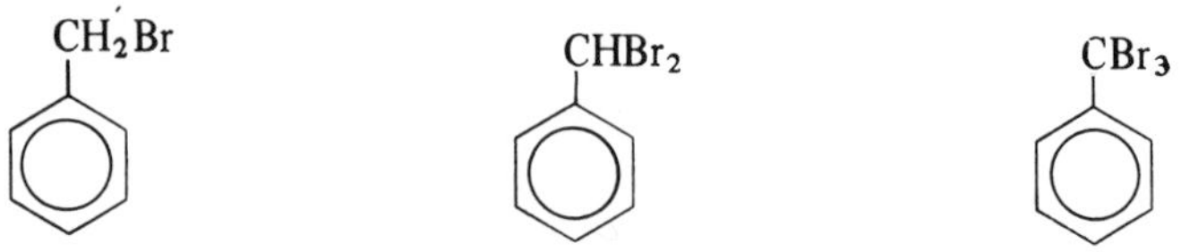

(bromomethyl) benzene (dibromomethyl) benzene (tribromomethyl) benzene

Halogenomethylbenzenes, $C_6H_4HalCH_3$

Methylbenzene reacts with chlorine or bromine in the presence of a halogen carrier, forming a mixture of the 1-chloro-2- and 1-chloro-4-methylbenzenes:

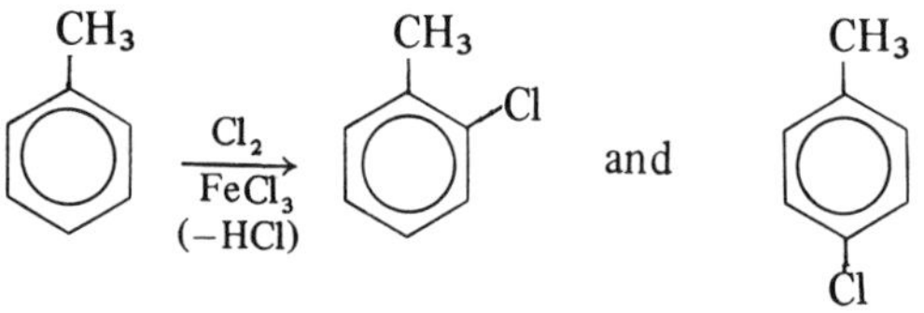

$$C_6H_5CH_3 \xrightarrow[FeCl_3\ (-HCl)]{Cl_2} \text{1-chloro-2-methylbenzene and 1-chloro-4-methylbenzene}$$

The chloromethylbenzenes are liquids similar to halogenobenzenes.

The halogen atom, like that in halogenobenzenes, is unreactive. Alkalis, potassium cyanide and ammonia solution do not react with them (contrast the halogenoalkanes and (chloromethyl)benzenes (p. 108)). Potassium manganate(VII) or nitric acid oxidizes the $-CH_3$ side-chain to $-COOH$:

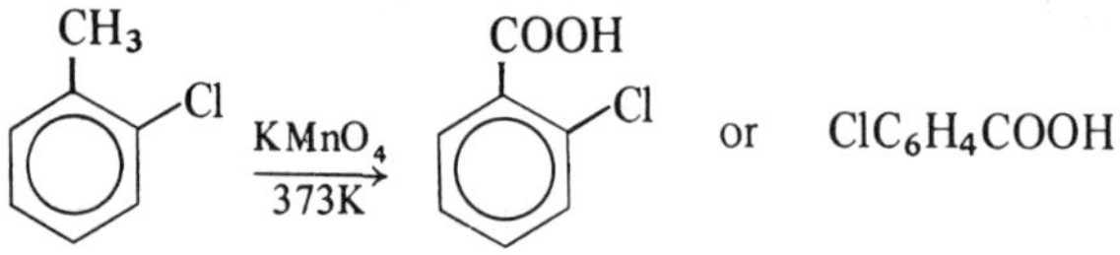

$$ClC_6H_4CH_3 \xrightarrow[373K]{KMnO_4} ClC_6H_4COOH$$

chlorobenzoic acid

Properties of (chloromethyl) benzene

It is a colourless liquid with a pungent irritating smell. It boils at 452 K and is almost insoluble in water.

The halogen atom is reactive and therefore the chemical properties resemble those of halogenoalkanes (p. 96). For example, the chlorine atom is substituted when (chloromethyl)benzene reacts with (a) boiling water or *aqueous* alkali or moist silver oxide, (b) warm alcoholic potassium cyanide, and (c) warm alcoholic ammonia solution:

$$C_6H_5CH_2Cl \xrightarrow{H_2O \text{ or alkali}} C_6H_5CH_2OH$$
phenylmethanol

$$C_6H_5CH_2Cl \xrightarrow{KCN} C_6H_5CH_2CN$$
phenylethanenitrile

$$C_6H_5CH_2Cl \xrightarrow{NH_3} C_6H_5CH_2NH_2$$
(phenylmethyl)amine

Warm *alcoholic* alkali eliminates the elements of hydrogen halide from halogen compounds with two or more carbon atoms in the side-chain, e.g.

$$C_6H_5CH_2CH_2Cl \xrightarrow[\text{(alc)}]{KOH} C_6H_5CH{=}CH_2$$
phenylethene

(Chloromethyl)benzene can be nitrated and sulphonated, but oxidation to benzoic acid also occurs during nitration.

SUMMARY

Monohalogenobenzenes

C_6H_5Hal. Function group: C—Hal.

Preparation
Dry benzene with (a) $Cl_2(g)/FeCl_3$ or $AlCl_3$, or (b) $Br_2(l)/FeBr_3$. I_2 has no action.

Properties
Unlike monohalogenoalkanes they have no reaction with alkalis, KCN(alc), NH_3(alc), KOH(alc) or C_2H_5ONa.

Halogenomethylbenzenes

Four C_7H_7Br isomers:

$C_6H_5CH_2Br$ (chloromethyl)benzene and $C_6H_4BrCH_3$ 1-bromo-2-, 3- and 4-methylbenzenes

Comparison of chloroethane, chlorobenzene and (chloromethyl)benzene

Reagent	*Chloroethane*	*Chlorobenzene*	*(Chloromethyl)benzene*
KOH(aq)	Ethanol	No action	Phenylmethanol
KCN(alc)	Propanenitrile	No action	Phenylethanonitrile
NH_3(alc)	Ethylamine and $(C_2H_5)_2NH$ and $(C_2H_5)_3N$	No action	(Phenylmethyl)amine and other amines
Mg	C_2H_5MgCl	C_6H_5MgCl	$C_6H_5CH_2MgCl$
$AgNO_3$(aq)	No action	No action	No action
HNO_3	No action	Nitration to $O_2NC_6H_4Cl$	Nitration to $O_2NC_6H_4CH_2Cl$ (and oxidation)
H_2SO_4	No action	Sulphonation to $HOSO_2C_6H_4Cl$	Sulphonation to $HOSO_2C_6H_4CH_2Cl$
Li[AlH_4]	Ethane	Benzene	Methylbenzene
Cl_2	$C_2H_4Cl_2$, $C_2H_3Cl_3$, etc., to C_2Cl_6	$C_6H_4Cl_2$ (1,2-and 1,4-compounds), $C_6H_3Cl_3$, etc., to C_6Cl_6	$C_6H_5CHCl_2$ and $C_6H_5CCl_3$

QUESTIONS

1. Outline the laboratory preparation of bromobenzene. Mention four reactions of bromobenzene which differ from those of bromoethane.

2. Name and write displayed formulae of four isomers of molecular formula C_7H_7I. Mention briefly methods of preparing any three of these isomers.

3. Arrange the following halides in order of their rates of hydrolysis, putting the one with the fastest rate first:

C_2H_5Cl $\quad$ C_6H_5Cl $\quad$ $C_6H_5CH_2Cl$ $\quad$ $C_6H_5CH_2Br$

Describe concisely experiments by which you would attempt to confirm the order you have given.

4. An aromatic chlorine compound has a relative molecular mass of 195.5. 0.1 g of the compound yielded, on combustion in oxygen, 0.158 g of carbon dioxide and 0.023 g of water. What is the molecular formula of the compound?

5. An aromatic alcohol of formula $C_8H_{10}O$ forms phenylethene (styrene) when heated with concentrated sulphuric acid. Write two possible structural formulae for the alcohol, and suggest a mechanism for the reaction. Phenylethene combines with hydrogen bromide; write down the structure of the addition compound and the probable mechanism of the addition reaction. The addition compound reacts with alcoholic potassium hydroxide to form the aromatic alcohol and phenylethene. Write equations showing the concurrent formation of the two products and explain the functions of the alkali in the two reactions. (C.)

6. State briefly, giving equations, how you would substitute the halogen atom in 1-chlorobutane and in 1-bromo-2-methylbenzene with these groups: $-OH$, $-NH_2$ and $-CN$. Write equations showing the first product formed when bromine reacts with (a) benzene, (b) cyclohexene, and (c) cyclohexane.

7. Outline one chemical test, giving equations and conditions, to distinguish between (a) 2-chlorobutane and 2-bromobutane, (b) bromobenzene and bromocyclohexane, and (c) 1-bromobutane and 1-bromobutene.

Chapter 8
ALCOHOLS

Alcohols contain one or more hydroxyl groups (—OH) joined to a grouping of carbon and hydrogen only (but not to a benzene ring). Those with two and three hydroxyl groups per molecule are called *diols* and *triols* respectively. Monohydric alcohols which are derivatives of alkanes have the general formula $C_nH_{2n+1}OH$, and their functional group is $-CH_2OH$. The first two members of the series are methanol (methyl alcohol) CH_3OH and ethanol (ethyl alcohol or simply 'alcohol') C_2H_5OH or CH_3CH_2OH.

Name	*Formula*	*B.p./K*	*Density at 298 K/g cm^{-3}*	*Solubility/g per 100 g*
Methanol	CH_3OH	337.7	0.793	Infinite
Ethanol	CH_3CH_2OH	351.5	0.789	Infinite
Propan-1-ol	$CH_3CH_2CH_2OH$	370.4	0.804	Infinite
Butan-1-ol	$CH_3CH_2CH_2CH_2OH$	390.9	0.810	8.3
Pentan-1-ol	$CH_3CH_2CH_2CH_2CH_2OH$	411.2	0.815	2.6

There are two alcohol isomers C_3H_8O or C_3H_7OH:

$CH_3CH_2CH_2OH$

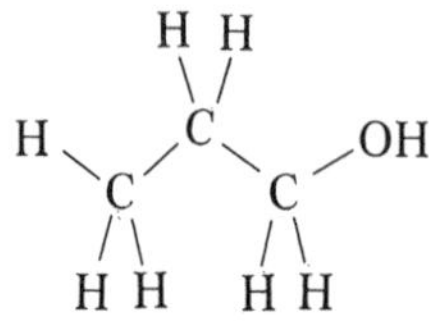

propan-1-ol

$(CH_3)_2CHOH$

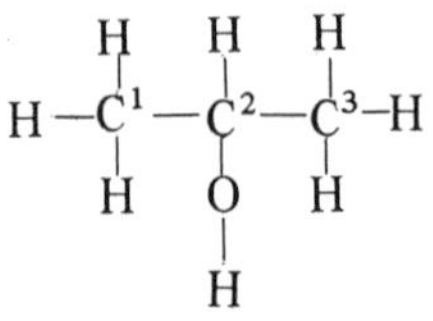

or

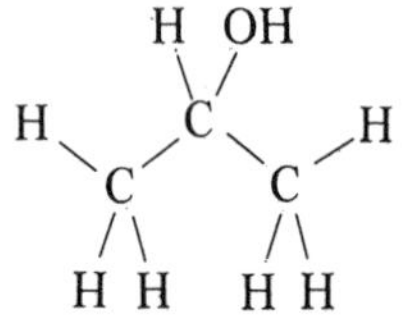

propan-2-ol

There are four isomeric alcohols C_4H_9OH:

$CH_3CH_2CH_2CH_2OH$ butan-1-ol

$CH_3CH_2CHOHCH_3$ butan-2-ol

$$\begin{array}{c} CH_3CHCH_2OH \\ | \\ CH_3 \end{array}$$
2-methylpropan-1-ol

$$(CH_3)_3COH \text{ or } \begin{array}{c} CH_3COHCH_3 \\ | \\ CH_3 \end{array}$$
2-methylpropan-2-ol

Methanol, ethanol and three of the above six isomers contain the monovalent $-CH_2OH$ group and are called *primary* alcohols. Propan-2-ol and butan-2-ol contain the divalent $>CHOH$ or $>\overset{\overset{H}{|}}{C}-OH$ group and are called *secondary* alcohols, and 2-methylpropan-2-ol contains the trivalent $\rightarrow C-OH$ group and is a *tertiary* alcohol.

Formation or preparation of primary alcohols

1. *Hydrolysis of halogenoalkanes.* Heat the halogenoalkane under reflux with *aqueous* sodium hydroxide, potassium hydroxide or 'moist silver oxide' (p. 96):

$$R-Hal(l) + OH^-(aq) \rightarrow Hal^- + R-OH$$

2. *Hydration of an alkene.* Addition of sulphuric acid to alkenes, occurs at 373 K and water decomposes the product:

$$>C=C< \xrightarrow{H_2SO_4} >\underset{H}{\underset{|}{C}}-\underset{OSO_2OH}{\underset{|}{C}}< \xrightarrow{H_2O} >\underset{H}{\underset{|}{C}}-\underset{OH}{\underset{|}{C}}<$$

Ethene and steam combine under pressure at 600 K with phosphoric(V) acid as a catalyst (p. 48).

Other methods include the reduction of aldehydes (p. 154) the action of nitrous acid, HNO_2, on primary amines (p. 235) and the reduction of carboxylic acids or their esters (p. 213).

To prepare ethanol

The enzyme zymase in yeast converts glucose to ethanol and carbon dioxide:

$$C_6H_{12}O_6 \rightarrow 2CO_2 + 2C_2H_5OH$$

The enzyme acts as a catalyst. The change is called *fermentation.*

Dissolve glucose (about 150 g) in about 500 cm^3 of hot water. Add ammonium phosphate (1 g) and potassium nitrate (1 g) to provide food for the yeast cells. Add $1\frac{1}{2}$ litres of cold water. Make yeast (50 g) into a thin paste with water and add to the glucose solution. Allow the mixture to stand at about 303 K for at least two days. Bubbles of carbon dioxide are evolved. The product is dilute aqueous ethanol. Use fractional distillation to concentrate the solution.

Reactions of methanol and ethanol

1. *Combustion.* Add a few drops of the alcohols to separate watch-glasses and burn them. Note the colour of the flames.

2. *Sodium or potassium.* Add about 1 cm^3 of the pure dry alcohols to separate test-tubes, and then add small pieces of freshly cut sodium or potassium. Ignite the gas produced. Pour the liquid products into small evaporating dishes and evaporate to dryness on a water bath. Observe the solid residues. Find the action of the residues on litmus.

3. *Phosphorus pentachloride.* To about 2 cm^3 of the pure dry alcohols add small pieces of phosphorus pentachloride. Observe any reaction.

4. *Ethanoic acid.* To 2 cm^3 of ethanol in a boiling tube add 1 cm^3 of glacial ethanoic acid and then 3 drops of concentrated sulphuric acid. Warm gently for 5 minutes. Add the product to a beaker containing aqueous sodium carbonate, which neutralizes the unchanged acids. Stir well and smell the product. Repeat the test with methanol.

5. *Oxidation.*
(a) Add about 10 cm^3 of methanol to an evaporating basin. Heat copper foil (about 2 cm square) or a copper coin until it is red-hot and drop it into the methanol. Note the change in appearance of the copper and observe the smell of the product.

(b) Add about 2 cm^3 of ethanol to about 2 cm^3 of saturated aqueous sodium (or potassium) dichromate(VI), $Na_2Cr_2O_7$. Carefully add about 2 cm^3 of concentrated sulphuric acid to the mixture. Warm gently. Note any colour change and smell the product.

(c) Repeat test (b) under more vigorous oxidizing conditions by using 4 cm^3 of dichromate solution and also of acid.

6. *Trihalomethane (haloform) reaction.* Do one of the following tests.
(a) Dissolve about 2.5 g of sodium carbonate crystals in the smallest volume of hot water. Cool the solution. Add about 1 cm^3 of ethanol and then 2 g of finely

powdered iodine a little at a time. Warm slowly to about 370 K. Observe the reaction. Find if methanol gives the reaction.

(b) Add 1 g of potassium iodide to about 2 cm^3 of ethanol. Add aqueous sodium chlorate(I), NaClO, until no more precipitate is formed. No heating is required.

Properties of methanol and ethanol

Physical properties. They are colourless liquids with a burning taste and pleasant odour, and are completely miscible with water. Ethanol freezes at 159 K and is therefore used in meteorological thermometers. Methanol is poisonous. Ethanol affects the brain and nervous system, and large quantities cause unconsciousness and even death.

The rise in boiling point which occurs in ascending the alcohol series is due to the greater number of electrons per molecule and therefore greater van der Waals' forces between molecules. It is not due merely to increase in mass.

Chemical properties

The reactions of ethanol and other primary alcohols can be classified as those of their —OH group, oxidation, dehydration, and trihalomethane reactions.

1. *Reactions of the hydroxyl group.*
(a) *Sodium or potassium.* The alkali metals react well but not violently, and they do not ignite and sometimes explode as in water. Sodium and potassium alkoxides are formed. Sodium ethoxide and methoxide are white deliquescent solids which turn litmus blue:

$$2C_2H_5OH(l) + 2Na(s) \rightarrow H_2(g) + \underset{\text{sodium ethoxide}}{2C_2H_5O^-Na^+}$$

$$2ROH + 2K \rightarrow H_2 + \underset{\text{potassium alkoxide}}{2RO^-K^+}$$

Compare:

$$2HOH + 2Na \rightarrow H_2 + \underset{\text{sodium hydroxide}}{2HO^-Na^+}$$

(b) *Phosphorus pentahalides.* The pentachloride reacts vigorously to form chloroalkanes. Hydrogen chloride fumes are evolved, and this is a test for the presence of an —OH group in aliphatic compounds.

$$C_2H_5OH(l) + PCl_5(s) \rightarrow HCl(g) + PCl_3O + C_2H_5Cl$$

The reactions of alcohols with phosphorus and bromine or iodine, phosphorus trichloride and sulphur dichloride oxide are on p. 91. The reactions are used to prepare halogenoalkanes.

(c) *Acids* (formation of esters). An alcohol and acid react reversibly to form an ester and water:

$$ROH + HA \rightleftharpoons H_2O + RA$$

$$C_2H_5OH(l) + H_2SO_4(l) \rightleftharpoons H_2O + \underset{\text{ethyl hydrogensulphate}}{C_2H_5HSO_4(aq)}$$

Strong acids react fairly readily but weak acids such as ethanoic acid react very slowly and the reaction is catalysed by oxonium ions H_3O^+ (from a strong acid), p. 116.

$$CH_3CO\boxed{OH + H}OC_2H_5 \rightleftharpoons H_2O + \underset{\text{ethyl ethanoate}}{CH_3COOC_2H_5}$$

Ethanol containing the isotope ^{18}O reacts with ethanoic acid. The ^{18}O is found in the ester and not in the water, and this fact proves that the alkyl-oxygen bond of the alcohol does not break.

Diethyl sulphate is formed by the action of sulphuric acid at low temperatures. It is removed by vacuum distillation. The low temperature avoids formation of ethoxyethane and ethene by dehydration:

$$2C_2H_5OH(l) + H_2SO_4(l) \rightleftharpoons H_2O + (C_2H_5)_2SO_4(l)$$

2. *Oxidation of alcohols.*
(a) Alcohols burn in air or oxygen with a faint blue flame, forming carbon dioxide and water:

$$C_2H_5OH + 3O_2 \rightarrow 2CO_2 + 3H_2O$$

(b) Less vigorous oxidizing agents convert primary alcohols to aldehydes. Hot copper acts as a dehydrogenation catalyst for the reactions:

$$CH_3OH(g) \xrightarrow{Cu} H_2 + \underset{\text{methanal}}{HCHO(g)}$$

$$C_2H_5OH(g) \xrightarrow{Cu} H_2 + \underset{\text{ethanal}}{CH_3CHO(g)}$$

The aldehydes are also formed when alcohol vapour mixed with air is passed over hot copper gauze:

$$2C_2H_5OH(g) + O_2 \rightarrow 2H_2O + 2CH_3CHO(g)$$

A mixture of sodium or potassium dichromate(VI), $Na_2Cr_2O_7$, and sulphuric acid is the usual oxidizing agent used, and the golden yellow mixture is turned to a green chromium(III) salt.

(In the 'breathalyser' test, if the special orange crystals turn blue, the breath contains much ethanol vapour.)

$$3RCHO(l) + Cr_2O_7^{2-}(aq) + 8H^+(aq) \rightarrow 3RCOOH + 2Cr^{3+}(aq) + 4H_2O$$

(c) If excess of the dichromate–acid mixture is used, or if the aldehyde remains in contact with the oxidizing agent, further oxidation to a carboxylic acid occurs:

$$CH_3OH \xrightarrow{Cr_2O_7^{2-}/H^+} HCHO \rightarrow \underset{\text{methanoic acid}}{HCOOH}$$

$$C_2H_5OH \xrightarrow{Cr_2O_7^{2-}/H^+} CH_3CHO \rightarrow \underset{\text{ethanoic acid}}{CH_3COOH}$$

The acid produced contains all the carbon atoms of the alcohol used and none is converted to carbon dioxide.

$$\underset{\text{primary alcohol}}{R-CH_2-OH} \rightarrow \underset{\text{aldehyde}}{R-C\begin{matrix}\diagup H \\ \diagdown\!\!\diagdown O\end{matrix}} \rightarrow \underset{\text{carboxylic acid}}{R-C\begin{matrix}\diagup OH \\ \diagdown\!\!\diagdown O\end{matrix}}$$

3. *Dehydration of alcohols.*

(a) Excess concentrated sulphuric acid at 440 K or phosphoric(V) acid at 470 K dehydrates ethanol to ethene; dehydration of ethanol also occurs when ethanol vapour is passed over aluminium oxide at 670 K or hot pumice stone or porcelain:

$$CH_3CH_2OH \xrightarrow[440\ K]{\text{conc } H_2SO_4\text{(excess)}} H_2C{=}CH_2$$

Methanol cannot be dehydrated in this way because its molecule contains only one carbon atom and it is not possible to form a carbon–carbon double bond.

(b) Dehydration by concentrated sulphuric acid at 410 K with ethanol in excess produces ethoxyethane (ether):

$$2C_2H_5OH\text{(excess)} \xrightarrow[410\ K]{\text{conc } H_2SO_4} C_2H_5OC_2H_5$$

The mechanism is discussed on p. 116.

4. *Trihalomethane (haloform) reaction.* Ethanol (but not methanol) reacts with sodium chlorate(I), NaClO, or bleaching powder to form trichloromethane, $CHCl_3$. Ethanol reacts with bromine and alkali to form tribromomethane, $CHBr_3$, and with iodine and alkali to form yellow solid triiodomethane CHI_3 (p. 112). The *triiodomethane* or *iodoform* test is given by all substances containing the groups CH_3CHOH- or CH_3CO-, e.g. ethanol, propan-2-ol and propanone (p. 158).

5. *Acid chlorides and anhydrides.* These form esters, e.g. $CH_3COOC_2H_5$ and $C_6H_5COOC_2H_5$ (p. 197).

Protonation of alcohols

Protons play an important part in some reactions of alcohols, for example, with hydrogen halides and sulphuric acid. Alcohols are covalent compounds and hardly form any hydroxide ions by dissociation. In the presence of acid, a proton joins to the oxygen atom of an alcohol such as ethanol:

$$CH_3CH_2{-}O{-}H \xrightarrow{acid} CH_3CH_3{-}\overset{+}{O}\begin{matrix} \diagup H \\ \diagdown H \end{matrix} \quad \text{or} \quad CH_3CH_2\overset{+}{O}H_2$$

The product, *protonated ethanol,* is sometimes called the ethyloxonium ion (an alkyloxonium ion has a positive charge on oxygen).

Nucleophiles such as Cl^-, Br^- and I^- (from hydrogen halides) react readily with protonated ethanol, which can lose water much more readily than unprotonated ethanol can lose an OH group:

$$C_2H_5OH \xrightarrow{H^+} C_2H_5\overset{+}{O}H_2 \xrightarrow{Br^-} C_2H_5Br + H_2O$$

Concentrated sulphuric acid and excess ethanol form ethoxyethane (ether). Some ethanol is protonated and then other unchanged ethanol reacts as a nucleophile:

$$C_2H_5OH(l) + C_2H_5\overset{+}{O}H_2(l) \rightarrow H_3O^+ + C_2H_5OC_2H_5(l)$$

Protons catalyse esterification reactions between an alcohol and a carboxylic acid, e.g.:

$$CH_3COOH(l) + HOC_2H_5(l) \rightleftharpoons CH_3COOC_2H_5(l) + H_2O$$

Reaction is very slow unless concentrated sulphuric acid or hydrogen chloride is added to the mixture. The ethanoic acid is protonated and then ethanol acts as a nucleophile:

$$CH_3CO\overset{+}{O}H_2(l) + C_2H_5OH(l) \rightleftharpoons CH_3COOC_2H_5(l) + H_3O^+$$

H_2O is lost from the protonated ethanoic acid, and therefore the alkyl–oxygen bond of the ethanol is not broken (p. 114).

Hydrogen bonds in alcohols

Alcohols resemble water in having fairly large dipole moments:

	Water	Methanol	Ethanol	Propan-1-ol
Dipole moment/10^{-30} C m	6.3	5.7	5.7	5.7

The boiling points of water and alcohols are much higher than those of the corresponding sulphur compounds (sulphur and oxygen are both in group 6 of the Periodic Table):

	H_2O	H_2S	C_2H_5OH	C_2H_5SH
B.p./K	373	213	351	310

The anomalous properties of alcohols and of water are due to the presence between their molecules of hydrogen bonds. A hydrogen bond is formed when a hydrogen atom which is already joined to an oxygen (or nitrogen or fluorine) atom joins with another of these atoms. Oxygen is strongly electronegative and its atom has a small radius. The oxygen atom in an alcohol molecule attracts the electron pair of the covalent bond and forms a strongly polar molecule: $C_2H_5-\overset{\delta-}{O}-\overset{\delta+}{H}$. The hydrogen atom with a small positive charge then joins electrostatically to another oxygen atom and forms a hydrogen bond, usually represented by four dots:

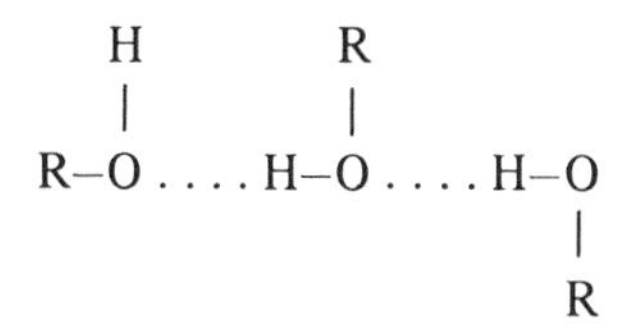

The lower alcohols are soluble in water because their polar molecules form links with the polar molecules of water. Higher alcohols are less soluble or insoluble in water, e.g. only 0.1 g of heptanol $C_7H_{15}OH$ dissolves in 100 g of water; the large hydrocarbon group in the higher alcohols accounts for their lower solubilities. Functional groups such as OH and NH_2 containing oxygen and nitrogen cause water solubility in molecules with less than six carbon atoms.

Ionic compounds dissolve in water, methanol, ethanol and lower alcohols. The polar water or alcohol molecules form electrostatic bonds with ions. For example, the positive hydrogen atom of an alcohol molecule joins to a negative ion and the negative oxygen atom of the alcohol joins to a positive ion. In other words, ionic compounds are solvated by alcohol just as they are hydrated by water. The existence of the compounds $CaCl_2 \cdot 3C_2H_5OH$ and $CaCl_2 \cdot 4CH_3OH$ explains why methanol and ethanol cannot be dried with calcium chloride.

Displayed (graphic) formula of ethanol

Qualitative analysis shows that only carbon, hydrogen and oxygen are present. Quantitative analysis shows that the empirical formula is C_2H_6O. The vapour density of ethanol vapour is 23 and therefore the relative molecular mass is $2 \times 23 = 46$. The molecular formula is C_2H_6O $(24 + 6 + 16 = 46)$. If the valencies of carbon, hydrogen and oxygen are 4, 1 and 2 respectively, two structural formulae are possible:

```
    H   H                  H       H
    |   |                  |       |
H—C—C—O—H             H—C—O—C—H
    |   |                  |       |
    H   H                  H       H
```

When sodium or potassium reacts with ethanol, only one-sixth of the hydrogen is displaced and the product is C_2H_5ONa or C_2H_5OK. The reaction shows that one hydrogen atom is different from the others. The formula may be written $(C_2H_5O)H$.

Phosphorus pentachloride reacts with ethanol to form C_2H_5Cl. One chlorine atom replaces one oxygen and one hydrogen atom. This reaction shows that an ethanol molecule contains a hydroxyl group. The formula on the left above is therefore correct.

Ethanol is oxidized to ethanoic acid. Two hydrogen atoms are replaced by one oxygen atom. This reaction indicates the presence of a $-CH_2-$ group in ethanol. Ethanoic acid contains a methyl group (p. 185), and therefore this group must be present in ethanol because it would not be created by oxidation.

Uses of alcohols

The lower alcohols are solvents in the manufacture of lacquers, varnishes, stains, medicines, perfumes, and so on. They are raw materials for the manufacture of many organic compounds: for example, methanol is oxidized to methanal, used in making certain plastics, and ethanol is used to make ethoxyethane, trichloromethane, propanone, ethanoic acid, dyes, and synthetic rubber. Methanol and ethanol are used in some anti-freeze mixtures added to car radiators to prevent ice forming in cold weather. Ethanol is added to some petrols and it is also used as a fuel in domestic heaters. Beers, wines, whisky, brandy, gin and so on contain between 3 and 50 per cent of ethanol.

Manufacture of methanol

Synthesis. Carbon monoxide and hydrogen are passed at about 250 atm pressure and 620 K over a catalyst of zinc oxide and chromium(III) oxide:

$$CO(g) + 2H_2(g) \xrightarrow{ZnO,\ Cr_2O_3} CH_3OH(l)$$

The hydrogen–carbon monoxide mixture is obtained by the action of steam on natural gas or naphtha with a nickel catalyst at high temperature and pressure:

$$CH_4(g) + H_2O(g) \rightarrow CO(g) + 3H_2(g)$$

From wood. Sawdust and wood chippings are heated to about 600 K and the vapours evolved are cooled. The liquid which condenses consists of an aqueous layer and tar. The aqueous layer contains about 6 per cent of methanol together with propanone and ethanoic acid. A mixture of methanol and propanone obtained by distillation of the aqueous layer is used as a solvent. Methanol is sometimes called *wood spirit.* Only a small proportion of methanol is now produced from wood. The solid residue is carbon (wood charcoal).

Manufacture of ethanol

From ethene. Direct hydration of ethene occurs when it combines with steam at 600 K, 70 atm pressure, and with phosphoric(V) acid as catalyst:

$$C_2H_4 + H_2O \rightarrow CH_3CH_2OH$$

Indirect hydration is brought about by absorbing ethene in concentrated sulphuric acid at 373 K and then adding water to the ethyl hydrogensulphate formed:

$$CH_2{=}CH_2 \xrightarrow{H_2SO_4} C_2H_5HSO_4 \xrightarrow{H_2O} C_2H_5OH + H_2SO_4$$

Ethanol distils over when the mixture is heated.

(Propan-2-ol is obtained by direct and indirect hydration of propene, $CH_3CH{=}CH_2$.)

From sucrose. Sugar cane and sugar beet contain sucrose, $C_{12}H_{22}O_{11}$, which is extracted by soaking them in hot water. The solution is concentrated by boiling and then allowed to crystallize. The final mother liquor, called molasses, contains about 50 per cent of sucrose. The liquor is diluted and special yeast is added. The yeast produces complex organic compounds called enzymes, which catalyse the change of sucrose to ethanol:

$$C_{12}H_{22}O_{11}(aq) \xrightarrow[(+H_2O)]{\text{invertase}} \underset{\text{glucose}}{C_6H_{12}O_6(aq)} + \underset{\text{fructose}}{C_6H_{12}O_6(aq)}$$

$$C_6H_{12}O_6(aq) \xrightarrow{\text{zymase}} 2C_2H_5OH(aq) + 2CO_2(g)$$

Glucose and fructose are isomers. The change of sugars to ethanol by the action of enzymes in yeast is called *fermentation.*

From starch. Potatoes, barley, maize and rice are sources of starch. Potatoes are heated with steam under pressure. The tough cell walls break and the starch goes into solution. Malt (partly germinated barley) is added and the enzyme diastase in it converts the starch to maltose (an isomer of sucrose):

$$2(C_6H_{10}O_5)_n(aq) + nH_2O \xrightarrow{\text{diastase}} nC_{12}H_{22}O_{11}(aq)$$

Yeast is added, and the enzyme maltase converts maltose to glucose:

$$C_{12}H_{22}O_{11}(aq) + H_2O \xrightarrow{\text{maltase}} 2C_6H_{12}O_6(aq)$$

Zymase converts the glucose to ethanol.

The solution obtained by fermentation of sugar or starch contains about 8 per cent of ethanol. Fractionation produces *rectified spirit*, which contains 95.6 per cent ethanol by mass and 4.4 per cent water. This is a constant-boiling mixture and the water cannot be removed by distillation.

Anhydrous ethanol, called *absolute alcohol*, is made by refluxing rectified spirit with dry calcium oxide. The oxide removes most of the water. The last traces of water are removed by calcium metal.

Industrial methylated spirit contains by mass about 85 per cent ethanol, 11 per cent water, and 4 per cent of methanol, which makes it unfit for drinking. A blue dye is sometimes added to methylated spirit to distinguish it from water.

Surgical spirit is a colourless liquid which contains about 90 per cent ethanol, 5 per cent methanol, and water.

Beer, spirits and wines

Beer contains from 3 to 6 per cent ethanol. It is made from barley which is steeped in water until it has partly germinated, during which time the enzyme diastase is formed. The barley plant is now killed by heat treatment and the grain is dried. The dry partly germinated barley grain, called malt, is steeped in water at 333 K, and diastase converts starch to maltose. The product is boiled with hops (dried hop-plant flowers) to give it a bitter taste, cooled, and fermented with yeast at about 290 K. Colouring and flavouring substances are added.

Whisky is made by a similar process but the fermentation of the malt and maltose proceeds for longer periods to produce 10 to 15 per cent ethanol. Distillation produces whisky, with about 40 per cent ethanol.

Rum is made by fermentation of molasses, which is 50 per cent sucrose.

Wines, both red and white, are made from grapes. Juice pressed from the grapes contains glucose and fructose. The juice ferments naturally because yeast is on the skins of grapes.

Ports and *sherries* are made by adding ethanol to wines.

Brandy is obtained by fractionation of certain wines.

Cider is made by fermenting apple juice.

Ethane-1,2-diol, CH_2OHCH_2OH

Ethane-1,2-diol is the commonest diol. Its molecule contains two primary alcohol groups $-CH_2OH$. It is used for the manufacture of Terylene, in anti-freeze mixtures for car and aircraft engines, and to prevent ice forming on the wings of aircraft.

Ethane-1,2-diol is manufactured by hydrolysis of epoxyethane, $(CH_2)_2O$, with hot water (p. 48).

Physical properties. Ethane-1,2-diol is a colourless, odourless, hygroscopic liquid. Unlike ethanol, it has a sweet taste, a high boiling point (470.5 K), is denser than water (1.12 g cm^{-3}), and is very poisonous.

Chemical properties. The reactions with sodium, potassium, phosphorus halides, sulphur dichloride oxide and acids resemble those of monohydric alcohols.

$$\begin{matrix} CH_2OH \\ | \\ CH_2OH \end{matrix} \xrightarrow{Na} \begin{matrix} CH_2ONa \\ | \\ CH_2OH \end{matrix} \quad \text{then} \quad \begin{matrix} CH_2ONa \\ | \\ CH_2ONa \end{matrix}$$

Hydrogen is evolved. The sodium salts resemble sodium ethoxide.

$$CH_2OHCH_2OH + 2PCl_5 \rightarrow CH_2ClCH_2Cl + 2PCl_3O + 2HCl$$

This reaction shows that the alcohol contains two hydroxyl groups. The product is 1,2-dichloroethane, which has a symmetrical structure (p. 98) and therefore the two hydroxyl groups in the diol are attached to different carbon atoms. Ethane-1,2-diol and phosphorus pentabromide react similarly.

Hydrogen chloride at 373 K forms CH_2ClCH_2OH and then forms the dichloro-compound at higher temperatures.

Differences between secondary and tertiary alcohols

On oxidation, a primary alcohol yields first an aldehyde and then an acid. The acid contains the same number of carbon atoms per molecule as the alcohol, p. 114.

A secondary alcohol oxidizes first to a ketone and then, by vigorous oxidation, to an acid with fewer carbon atoms per molecule than the original alcohol. A secondary alcohol also forms a ketone when it is dehydrogenated by passing over heated copper, whereas a primary alcohol yields an aldehyde

$$(CH_3)_2CHOH \xrightarrow[\text{mild}]{[O]} (CH_3)_2C{=}O \xrightarrow[\text{vigorous}]{[O]} CH_3COOH + CO_2$$

propan-2-ol — propanone — propanoic acid

A tertiary alcohol is not easily oxidized but it breaks down under vigorous oxidation to form a ketone and then one or more acids. The ketone and each acid have fewer carbon atoms per molecule than the original alcohol.

$$(CH_3)_3COH \xrightarrow[\text{vigorous}]{[O]} (CH_3)_2C{=}O + CO_2 \xrightarrow{[O]} CH_3COOH + 2CO_2$$

A tertiary alcohol yields an alkene and water when its vapour is passed over hot copper:

$$(CH_3)_3COH \xrightarrow[600\ K]{Cu} H_2O + (CH_3)_2C{=}CH_2,$$

2-methylpropene

SUMMARY

Alcohols

R—OH (R = alkyl group). Functional groups: primary, $-CH_2OH$; secondary $>CHOH$; tertiary $\geqslant C-OH$. Diols and triols have 2 and 3 —OH groups:

CH_2OHCH_2OH ethane-1,2-diol

$CH_2OHCHOHCH_2OH$ propane-1,2,3-triol

Properties
(of C_2H_5OH or CH_3CH_2OH, ethanol)

Reagent	Product
Reactions of −OH group	
Na (or K)	$H_2(g) + C_2H_5O^-Na^+$, or $C_2H_5O^-K^+$ sodium ethoxide
PCl_5 or SCl_2O	C_2H_5Cl chloroethane
Red P + Br_2 (or I_2)	C_2H_5Br or C_2H_5I bromoethane iodoethane
H_2SO_4(conc)	$C_2H_5HSO_4$ ethyl hydrogensulphate, an ester
$CH_3COOH/H^+(aq)$	$CH_3COOC_2H_5$ ethyl ethanoate (acetate), an ester
Oxidation	
Combustion	$CO_2 + H_2O$ in air. Faint blue flame
Hot Cu catalyst	$H_2 + CH_3CHO$ ethanal
$Cr_2O_7^{2-}/H^+$ or MnO_4^-/H^+	CH_3CHO, then CH_3COOH ethanoic acid
Dehydration	
Hot Al_2O_3 or porcelain	$CH_2{=}CH_2$ ethene
H_2SO_4(conc), 170 °C or H_3PO_4, 200 °C	$CH_2{=}CH_2$ (acid in excess)
H_2SO_4(conc), 140 °C	$C_2H_5OC_2H_5$ (ethanol in excess) ethoxyethane
Warm alkali + I_2 or Br_2 or Cl_2 (or NaClO)	CHI_3(s) (yellow) or $CHBr_3$ or $CHCl_3$ triiodomethane
CH_3COCl or $(CH_3CO)_2O$	$CH_3COOC_2H_5$ acylation ethyl ethanoate
C_6H_5COCl	$C_6H_5COOC_2H_5$ benzoylation ethyl benzoate

Manufacture
Methanol: Naphtha or natural gas with Ni catalyst → $CO + H_2$ (synthetic gas). $CO + 2H_2$ + special catalysts at high temperature and pressure → CH_3OH.

Ethanol: (a) Hydration of ethene by either steam (with catalyst, high temperature and pressure) or H_2SO_4(conc) at 100°C then water. $C_2H_4 + H_2O \rightarrow C_2H_5OH$

(b) Fermentation of glucose and sucrose (sugar cane and sugar beet) and starch (potatoes, barley, rice).

Ethane-1,2-diol, CH_2OHCH_2OH

Reactions are similar to those of ethanol, e.g. with Na, K, PCl_5, $P + Br_2$, SCl_2O and acids.

Manufacture

Ethene + O_2 with catalyst, high temperature and pressure → C_2H_4O, epoxyethane

then water → CH_2OHCH_2OH

Comparison of primary, secondary and tertiary alcohols

Reagent/ reaction	*Primary* CH_3CH_2OH *ethanol*	*Secondary* $(CH_3)_2CHOH$ *propan-2-ol*	*Tertiary* $(CH_3)_3COH$ *2-methylpropan-2-ol*
Dehydration Al_2O_3, pumice, porcelain	$CH_2{=}CH_2$ ethene	$CH_3CH{=}CH_2$ propene	$(CH_3)_2C{=}CH_2$ 2-methylpropene
H_2SO_4(conc)	Alkene slowly	Intermediate speed	Alkene fast
Oxidation $Cr_2O_7^{2-}/H^+$	Readily to CH_3CHO then CH_3COOH (same number of carbon atoms)	Readily to $(CH_3)_2CO$. More vigorous conditions to CH_3COOH (fewer carbon atoms)	Resists reaction. Vigorous conditions to $(CH_3)_2CO$, then CH_3COOH (fewer carbon atoms)

QUESTIONS

1. Describe the preparation of pure ethanol from glucose.

2. How, and under what conditions, does ethanol react with (a) sodium, (b) phosphorus pentachloride, and (c) sulphuric acid?

3. Describe a chemical method to distinguish between methanol and ethanol. Write equations for any reactions that occur.

4. Write the full displayed formulae of two alkanes C_4H_{10}. Write the formulae and names of four alcohols derived from these alkanes. Use these formulae to explain the difference between primary, secondary and tertiary alcohols.

5. Write the molecular and displayed formulae of ethoxyethane and butan-1-ol. State, with your reason, which of these two isomers you expect to have the higher boiling point. Why is the boiling point of C_2H_5OH higher than that of C_2H_5SH?

6. Describe what you would expect to observe in the following tests: (a) sodium is added to methanol, (b) ethanol and acidified sodium dichromate(VI), $Na_2Cr_2O_7$, are warmed gently, (c) hydrogen chloride is dissolved in a mixture of ethanol and glacial ethanoic acid, CH_3COOH, and the mixture is heated under reflux for some minutes.

7. Analysis shows that an unknown compound ***X*** contains only carbon, hydrogen and oxygen. A sample of ***X*** weighing 0.155 g was burned in oxygen and resulted in the formation of 0.220 g of carbon dioxide and 0.135 g of water. What is the empirical formula of ***X***? A sample of ***X*** of mass 0.225 g was vaporized completely at 483 K and 760 mmHg pressure. The volume measured under these conditions was 145.5 cm^3. A similar volume of oxygen, measured under identical conditions, weighs 0.120 g. What is the molecular formula of ***X***? Write three possible structural formulae for ***X***. Decide which formula best fits the following properties of ***X***: (a) It is miscible in all proportions with water and ethanol, (b) it reacts with sodium metal, producing hydrogen gas, (c) it reacts with phosphorus pentachloride, whereby 2 oxygen and 2 hydrogen atoms are replaced by 2 chlorine atoms, and (d) it is oxidized to ethanedioic acid $(COOH)_2$ by heating it with nitric acid. Write equations for reactions of ***X*** mentioned in (b), (c) and (d) and give a systematic name to ***X***. ($C = 12$, $H = 1$, $O = 16$.) (L.)

8. Bromoethane may be prepared in the laboratory by distilling a mixture of ethanol, potassium bromide and concentrated sulphuric acid. The bromoethane is collected under water in a suitable receiver. The distillate is transferred to a separating funnel and, after removal of the aqueous layer, it is shaken with dilute sodium carbonate solution. The bromoethane layer is then allowed to stand over anhydrous calcium chloride. The resulting liquid is distilled and the fraction boiling in the temperature range 308–313 K is collected. (a) Name the organic compound(s), apart from ethanol and bromoethane, which may be present in the distillation flask during the preparation. (b) Explain why the reaction mixture usually turns brown on heating. (c) Explain why the condenser is fitted with an adaptor dipping below the surface of the water in the receiver. (d) Why is the bromoethane layer allowed to stand over calcium chloride? (e) What is the main impurity in the product likely to be? (f) Show that the maximum yield of bromoethane from 4 g of ethanol is about 9.5 g. (L.)

Chapter 9
AROMATIC HYDROXY-COMPOUNDS

Phenols are a series of compounds each of which contains one or more hydroxyl groups attached directly to a benzene ring. The simplest one is phenol, C_6H_5OH, which is also called carbolic acid. Three methylphenols and four hydroxymethylbenzenes exist:

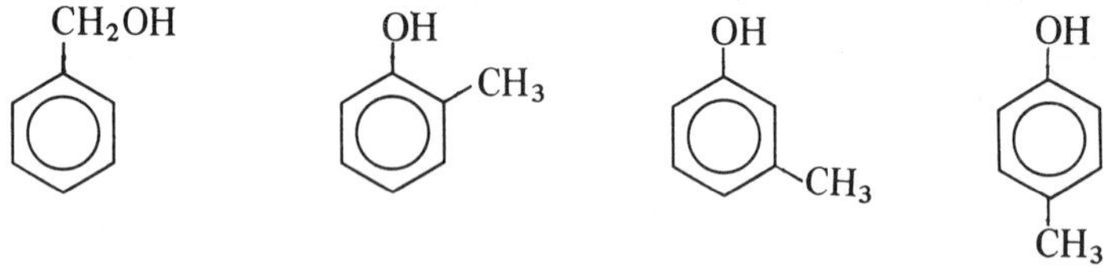

CH_2OH | OH, CH_3 | OH, CH_3 | OH, CH_3

phenylmethanol | 2- | 3- | 4-

methylphenols

The three isomeric phenols $C_6H_4(OH)_2$ are called benzene-1,2-diol, benzene-1,3-diol and benzene-1,4-diol.

Phenylmethanol, $C_6H_5CH_2OH$

This primary alcohol is the simplest aromatic or aryl alcohol. It is not a phenol.

Preparation

1. Boil (chloromethyl)benzene under reflux with aqueous sodium carbonate:

$$C_6H_5CH_2Cl \xrightarrow{\text{alkali}} C_6H_5CH_2OH + HCl \text{ (neutralized by alkali)}$$

2. *Cannizzaro reaction.* Benzaldehyde is treated with cold concentrated potassium hydroxide or sodium hydroxide:

$$2C_6H_5CHO(l) \xrightarrow[\text{NaOH}]{\text{KOH or}} C_6H_5COOK(aq) + C_6H_5CH_2OH(l)$$

In both preparations the alcohol is extracted with ether (p. 233), dried, and purified by distillation.

Physical properties. Phenylmethanol is a colourless liquid with a faint pleasant smell. It boils at 479 K. It is only slightly soluble in water.

Chemical properties. Many chemical reactions are similar to those of ethanol and other primary alcohols, and they include reactions with sodium, potassium, phosphorus halides, sulphur dichloride oxide, concentrated halogen acids, and oxidizing agents.

$$\xrightarrow{\text{Na}} H_2 + C_6H_5CH_2O^-Na^+$$

sodium phenylmethoxide

$$\xrightarrow{\text{PCl}_5} HCl + PCl_3O + C_6H_5CH_2Cl$$

(chloromethyl) benzene

$$\xrightarrow{\text{SCl}_2\text{O}} HCl(g) + SO_2(g) + C_6H_5CH_2Cl(l)$$

$$\xrightarrow{\text{HBr}} H_2O + C_6H_5CH_2Br$$

(bromomethyl) benzene

Dilute nitric acid or alkaline potassium manganate(VII) oxidizes it to benzaldehyde and then to benzoic acid:

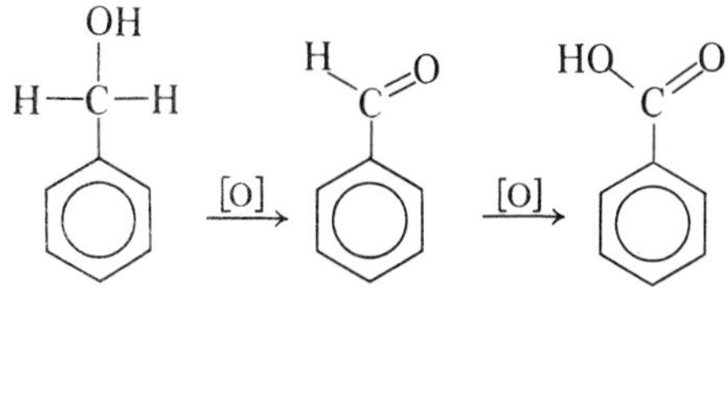

$$C_6H_5CH_2OH \xrightarrow{[O]} C_6H_5CHO \xrightarrow{[O]} C_6H_5COOH$$

Phenol, C_6H_5OH or C_6H_5–OH

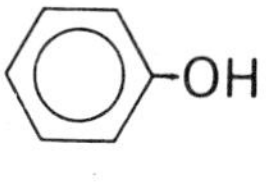

A dilute solution of phenol was the first antiseptic used in surgery, just over a century ago.

Manufacture

1. *From petroleum* (*cumene process*). Benzene and propene are obtained from crude oil by cracking. Their vapours are passed over phosphoric(V) acid, a catalyst, at moderate temperatures and pressures:

$$C_6H_6(g) + CH_3CH{=}CH_2(g) \rightarrow \underset{\substack{\text{(1-methylethyl) benzene} \\ \text{or cumene}}}{C_6H_5CH(CH_3)_2}$$

The product is oxidized by air and a catalyst to an unstable intermediate product, which reacts with dilute acid to form phenol and propanone:

$$C_6H_5CH(CH_3)_2 \rightarrow C_2H_5OH + (CH_3)_2CO$$

2. *From coal.* Fractionation of coal tar yields a fraction called middle oil, which contains phenol and methylphenols. Treatment with sodium hydroxide (forming C_6H_5ONa) and then carbon dioxide produces phenol and methylphenols.

3. *From benzene.* Convert benzene to benzenesulphonic acid, and fuse its sodium salt with sodium hydroxide:

$$C_6H_6 \rightarrow C_6H_5SO_2OH \rightarrow C_6H_5OH$$

Laboratory preparation. Phenol cannot be prepared either by introducing a hydroxyl group directly into a benzene ring or, under laboratory conditions, by replacing —Hal of a halogenobenzene with —OH.

1. Sulphonate benzene with hot concentrated sulphuric acid or cold fuming sulphuric acid and add the product to saturated aqueous sodium chloride. Sodium benzenesulphonate is precipitated:

$$C_6H_6(l) \xrightarrow{H_2SO_4} C_6H_5SO_2OH(aq) \xrightarrow{NaCl} C_6H_5SO_2ONa(s)$$

Fuse the sulphonate with sodium hydroxide at about 520 K in a nickel crucible:

$$C_6H_5SO_2ONa + 2NaOH \rightarrow H_2O + Na_2SO_3 + \underset{\text{sodium phenoxide}}{C_6H_5ONa}$$

Extract the sodium phenoxide from the cold mass with water. Acidify the solution to produce free phenol:

$$C_6H_6 \xrightarrow{H_2SO_4} C_6H_5SO_3H \xrightarrow{Na^+OH^-} C_6H_5O^-Na^+ \xrightarrow{H^+} C_6H_5OH$$

Reactions of phenol

1. Note the appearance and smell of the solid.

2. *Acidity.* Add a few phenol crystals to a boiling tube, half fill with water, and shake. Is phenol soluble? Add (a) litmus and (b) Universal Indicator to a few drops of the solution. Use the rest of the solution in the following tests. (The solution is corrosive.)

3. *Solubility.* To some solution add more solid phenol and shake. What happens? Warm the solution and note any changes.

4. *Alkalis.*
(a) To about 5 cm^3 of bench sodium carbonate solution add about 1 cm^3 of phenol. Is it soluble?

(b) Repeat the test using bench sodium hydroxide solution. Divide the product into two; pass carbon dioxide through one and add concentrated hydrochloric acid to the other.

5. *Bromine.* Add aqueous bromine drop by drop to aqueous phenol until a faint yellow colour persists. Note the colour of the precipitate. Filter (use a Buchner funnel and pump if necessary, Fig. 9.1) and recrystallize from an ethanol-water mixture.

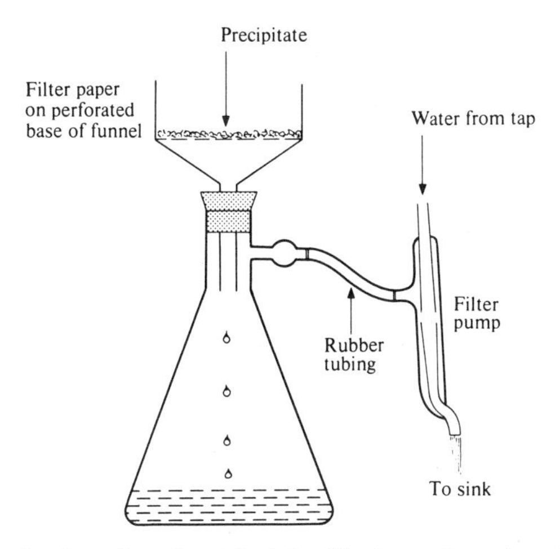

Fig. 9.1. Buchner funnel and flask for filtering under reduced pressure

7. *Benzoyl chloride.* Dissolve phenol (1 g) and sodium hydroxide (1.5 g) in about 20 cm^3 of water in a wide-mouthed bottle or flask. Add 2 cm^3 of benzoyl chloride (the liquid is dangerous and the vapour is irritating). Cork the bottle or flask and shake well for 10 minutes; remove the cork slightly from time to time

in order to release the pressure. Filter off the solid product. Recrystallize from ethanol, which should be heated on a water bath and not by a flame.

8. *Zinc.* Mix phenol (1 g) with about three times its volume of zinc powder in a dry test-tube. Heat the mixture. Identify the vapour formed by its smell and ignite it.

Properties of phenol

Physical properties. Phenol is a colourless crystalline hygroscopic solid which turns pink in light. Its odour is characteristic. It melts at 314 K, boils at 455 K, is completely miscible with water above 339 K, and is moderately soluble at ordinary temperatures. It is poisonous and can cause severe burns if the solid or concentrated solution is left on the skin.

Chemical properties. The reactions are those of the functional group, —OH, modified by the reactivity of the aromatic ring.

1. *Reactions of the hydroxyl group.*
(a) *Acidity.* Phenol is a weak acid and turns litmus pale red:

$$C_6H_5OH(aq) + H_2O \rightleftharpoons H_3O^+ + \underset{\text{phenoxide ion}}{C_6H_5O^-(aq)}$$

It forms salts with the hydroxides of sodium and potassium:

$$C_6H_5OH + NaOH \rightarrow H_2O + \underset{\text{sodium phenoxide}}{C_6H_5O^-Na^+}$$

Phenol is a weaker acid than carbonic acid; their dissociation constants are 10^{-10} and 3×10^{-7} mol dm^{-3} respectively. Therefore phenol does not react with sodium carbonate, and carbon dioxide liberates phenol from sodium phenoxide:

$$C_6H_5ONa + CO_2 + H_2O \rightarrow C_6H_5OH + NaHCO_3$$

Phenol is also a weaker acid than a carboxylic acid.

(b) *Acids, acid chlorides and acid anhydrides* (formation of esters). Phenol does not react with hydrogen halides or carboxylic acids (contrast ethanol). Phenol or a phenoxide form esters by reaction with an acid chloride or anhydride:

$$C_6H_5COCl(l) + NaOC_6H_5(s) \rightarrow NaCl(s) + \underset{\text{phenyl benzoate}}{C_6H_5COOC_6H_5(l)}$$

$$CH_3COCl(l) + HOC_6H_5(s) \rightarrow HCl(g) + \underset{\text{phenyl ethanoate}}{CH_3COOC_6H_5(l)}$$

The acid chlorides in the above equations are benzoyl chloride and ethanoyl (acetyl)chloride. Ethanoic (acetic) anhydride reacts as follows:

$$(CH_3CO)_2O(l) + HOC_6H_5(s) \rightarrow CH_3COOH(l) + CH_3COOC_6H_5(l)$$

The above reactions are examples of the *benzoylation* and *ethanoylation* (*acetylation*) of phenol (refer to p. 198).

(c) *Reduction.* Hot zinc dust reduces phenol to benzene:

$$C_6H_5OH(g \text{ or } l) + Zn(s) \rightarrow C_6H_6(g) + ZnO(s)$$

(d) *Phosphorus pentachloride.* Slight reaction occurs but little or no chlorobenzene is formed (contrast ethanol).

2. *Substitution reactions affecting the ring.* The hydroxyl group makes the benzene ring much more reactive than it is in benzene. Phenol reacts readily with chlorine, bromine, nitric acid and sulphuric acid. Substitution occurs in the 2- and 4-positions.

(a) *Halogenation.* Phenol and excess chlorine form 2,4,6-trichlorophenol. Chlorination of molten phenol is used commercially in the manufacture of 2,4-dichlorophenol, used to make selective weedkiller:

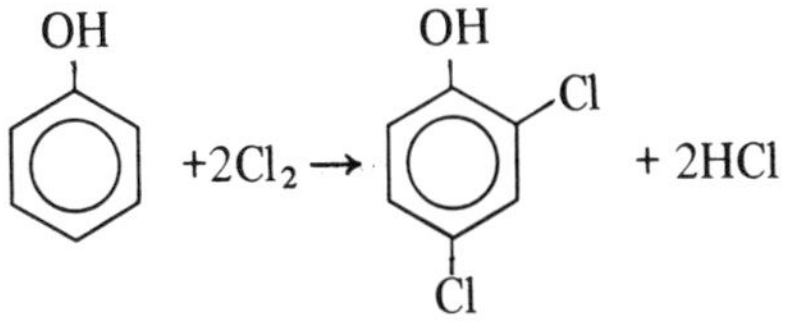

OH OH Cl Cl

$$+2Cl_2 \rightarrow \quad + 2HCl$$

A white precipitate of 2,4,6-tribromophenol is formed when bromine water is added to phenol:

$$C_6H_5OH(aq) + 3Br_2(aq) \rightarrow C_6H_2Br_3OH(s) + 3HBr(aq)$$

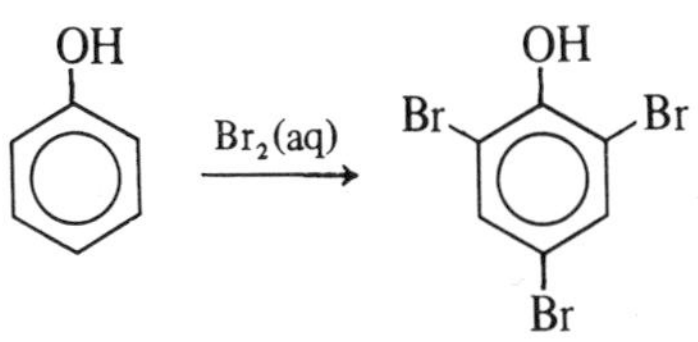

OH OH Br Br Br

$$\xrightarrow{Br_2(aq)}$$

(b) *Nitration.* Phenol and cold diluted nitric acid (25 per cent) readily form 2- and 4-nitrophenols as a dense oil (the pure substances are solids):

$$C_6H_5OH \xrightarrow[\text{cold}]{\text{dil } HNO_3} \text{2-}O_2NC_6H_4OH \text{ and } \text{4-}O_2NC_6H_4OH$$

A mixture of concentrated nitric and sulphuric acids at 373 K converts phenol to 2,4,6-trinitrophenol, which separates as yellow crystals when the reaction mixture is added to cold water:

$$C_6H_5OH \xrightarrow[\text{conc } H_2SO_4,\ 373\text{ K}]{\text{conc } HNO_3} \text{2,4,6-}(O_2N)_3C_6H_2OH$$

(c) *Sulphonation.* Concentrated sulphuric acid at 373 K forms 2- and 4-hydroxybenzenesulphonic acids, and further sulphonation yields 4-hydroxybenzene-1,3-disulphonic acid:

$$C_6H_5OH \xrightarrow[373\text{ K}]{H_2SO_4} \text{2-}HOSO_2C_6H_4OH \text{ and } \text{4-}HOSO_2C_6H_4OH$$

(d) *Reduction.* Hydrogen and phenol vapour combine when passed over a nickel catalyst at 470 K:

$$C_6H_5OH + 3H_2 \rightarrow \underset{\text{cyclohexanol}}{C_6H_{11}OH}$$

This alcohol is used in the manufacture of nylon.

3. *Iron*(*III*) *chloride*. The aqueous solution forms a violet colour with phenol, and most simple aromatic hydroxy-compounds give purple, violet or green colours. This is a test for phenols.

Uses of phenol

Phenol is a raw material used in the manufacture of plastics, nylon, explosives, and selective weedkillers. Under certain conditions, phenol and methanal react to form Bakelite, which is a plastic used to make electrical plugs and switches and small household articles. It is a tough plastic with low electrical conductivity. Phenol was once used, in 3 per cent solution, as an antiseptic and sterilizing agent, but more effective and less dangerous germicides are now available.

Delocalization of electrons in phenol

The usual formulae (given below) show that the oxygen atom in phenol and in ethanol has two lone pairs of electrons and in the phenoxide ion and the ethoxide ion it has three lone pairs:

$C_6H_5{-}\ddot{\underset{\cdot\cdot}{O}}{-}H$	$H{-}CH_2{-}CH_2{-}\ddot{\underset{\cdot\cdot}{O}}{-}H$	$C_6H_5{-}\ddot{\underset{\cdot\cdot}{O}}{:}^-$	$H{-}CH_2{-}CH_2{-}\ddot{\underset{\cdot\cdot}{O}}{:}^-$
phenol	ethanol	phenoxide ion	ethoxide ion

The carbon–oxygen bonds in phenol and ethanol seem to be the same, but this is not so. The bond length in ethanol and other alcohols is 0.143 nm, while that in phenol is only 0.136 nm. The bonds in the two ions also differ.

Delocalization of two of the electrons of the oxygen atom in the phenoxide ion accounts for the difference. Delocalization cannot occur in ethanol or the ethoxide ion, and all the electrons are localized in definite covalent bonds. Two electrons of the oxygen atom in phenol and the phenoxide ion overlap with the π-orbitals of the benzene ring carbon atoms. Delocalization makes the phenoxide ion more stable than the ethoxide ion and the phenoxide ion is less ready to combine with a hydrogen ion. In other words, the phenoxide ion is a weaker base and therefore phenol is more acidic than ethanol. Dissociation constants in mol dm^{-3} are:

Phenol 10^{-10}; 4-methylphenol 6.4×10^{-11}; methanol 10^{-16};

ethanol 1.26×10^{-16}

The following two formulae are made possible by electron delocalization:

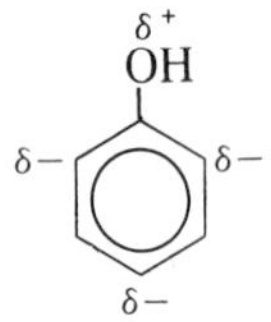

phenol

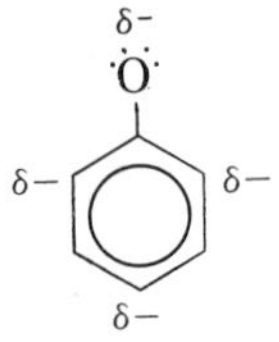

phenoxide ion

Slight negative charges appear in the 2- and 4-positions and therefore attack by electrophiles such as Br^+ and NO_2^+ is easy at those positions.

Only three carbon atoms in phenol and its ion can have a negative charge. It is not possible to draw correct formulae (those which obey the octet rule and in which electrons are paired) to show a negative charge on any of the other carbon atoms.

SUMMARY

Phenylmethanol, $C_6H_5CH_2OH$

Preparation

1. *Reflux.* (Chloromethyl)benzene, $C_6H_5CH_2Cl$, with alkali.

2. *Cannizzaro reaction.* Benzaldehyde, C_6H_5CHO, with NaOH or KOH (cold, conc) $\rightarrow C_6H_5CH_2OH + C_6H_5COONa$

Properties

Reacts as primary alcohol with Na, K, PCl_5, $P + Br_2$, $P + I_2$, SCl_2O, acids, and oxidizing agents.

Phenol, C_6H_5OH

Manufacture

1. Crude oil cracked $\rightarrow$ benzene + propene, $CH_3CH{=}CH_2$, and with catalyst these form (1-methylethyl)benzene, $C_6H_5CH(CH_3)_2$. Air + catalyst and then acid $\rightarrow$ phenol + propanone.

2. Coal tar fractionated to middle oil. NaOH and then $CO_2 \rightarrow$ phenol.

3. Benzene $\rightarrow C_6H_5SO_2OH$. Fuse with NaOH $\rightarrow$ phenol.

Comparison of ethanol, phenol and phenylmethanol

Reagent/reaction property	*Ethanol*	*Phenol*	*Phenylmethanol*
Solubility	Complete	Moderate	Slight
Litmus	Neutral	Faint red	Neutral
Na	C_2H_5ONa	C_6H_5ONa	$C_6H_5CH_2ONa$
NaOH	No action	C_6H_5ONa	No action
PCl_5, SCl_2O	C_2H_5Cl	Little action	$C_6H_5CH_2Cl$
HCl, HBr	C_2H_5Hal	No action	$C_6H_5CH_2Hal$
CH_3COOH	$CH_3COOC_2H_5$	No action	$CH_3COOCH_2C_6H_5$
CH_3COCl $(CH_3CO)_2O$	$CH_3COOC_2H_5$	$CH_3COOC_6H_5$	$CH_3COOCH_2C_6H_5$
Oxidation	CH_3CHO, then CH_3COOH	Difficult	C_6H_5CHO, then C_6H_5COOH
Dehydration	$C_2H_5OC_2H_5$	None	None
Sulphonation	None	$HOSO_2C_6H_4OH$	None
Nitration	None	$O_2NC_6H_4OH$	None
Cl_2	CCl_3CHO	$C_6H_2Cl_3OH$	
Br_2		$C_6H_2Br_3OH$	
$FeCl_3$	No action	Blue or purple	No action

QUESTIONS

1. Outline two methods of preparing phenylmethanol. What is the action on this compound of (a) sodium, (b) sulphur dichloride oxide, and (c) dilute nitric acid?

2. Mention two methods by which phenol may be obtained from benzene. What is the action on phenol of (a) water, (b) sodium hydroxide, (c) iron(III) chloride solution?

3. Compare and contrast the reactions of the —OH group in phenol and ethanol. Account for the differences.

4. Mention two chemical reactions which are given by phenol but not by phenylmethanol, and two chemical reactions given by phenylmethanol but not by phenol.

5. A student submitted the following account of the preparation and purification of a specimen of phenol. Point out and correct any errors you can find in it. 'Hydrated calcium benzenesulphonate was dissolved in water and boiled under reflux with calcium hydroxide for 12 hours. The resulting solution contained phenol, which was formed according to the equation:

$$C_6H_5SO_3Ca + CaOH \rightleftharpoons C_6H_5OH + Ca_2SO_3$$

The phenol was extracted once with a large volume of ether and the ether removed from the extract by distillation under reduced pressure over a bunsen burner. The liquid residue was dissolved in sodium carbonate solution, reprecipitated with dilute hydrochloric acid, separated, and redistilled under reduced pressure to give pure liquid phenol.' (L.)

6. The relative molecular mass of an organic acid ***A*** is 138. The percentage composition by mass of acid ***A*** is: carbon 60.87, hydrogen 4.35, and the rest is oxygen only. Acid ***A*** reacts with a solution of hydrogen chloride in ethanol when the mixture is boiled under reflux for some time, forming a compound ***B***. The molecular formula of ***B*** is $C_9H_{10}O_3$ ***B*** dissolves in a cold dilute solution of sodium hydrogencarbonate. Deduce the molecular formula of ***A***, suggest three possible structures for ***A***, and explain the formation of ***B***.

7. How does bromine react with each of the following groups of organic substances: (a) alkenes, (b) phenols, (c) benzene and methylbenzene? State the essential practical conditions and write an equation for each reaction you mention.

8. (a) Describe the manufacture of phenol from benzene by the cumene process and outline the industrial importance of phenol. (b) How, and under what conditions, would you expect a compound with the formula

OH

$CH{=}CHCH_2OH$

to react, if at all, with (i) sodium, (ii) sodium hydroxide, (iii) phosphorus pentachloride, (iv) ethanoic acid, (v) potassium manganate(VII)? (C.)

Chapter 10
ETHERS

The general formula of ethers is R′—O—R″, in which R′ and R″ are alkyl or aryl groups. If R′ and R″ are the same, the ether is a simple or symmetrical ether, and if they are different it is a mixed or unsymmetrical ether.

$$O\begin{matrix} \nearrow CH_3 \\ \searrow CH_3 \end{matrix} \quad \text{or} \quad (CH_3)_2O \quad \text{or} \quad CH_3OCH_3$$

methoxymethane

$$O\begin{matrix} \nearrow C_2H_5 \\ \searrow C_2H_5 \end{matrix} \quad \text{or} \quad (C_2H_5)_2O \quad \text{or} \quad C_2H_5OC_2H_5$$

ethoxyethane
(ether)

Methoxyethane, $CH_3OC_2H_5$, is an example of a mixed ether. The functional group of ethers is C—O—C. Ethers contain the polar $\overset{\delta+}{C}—\overset{\delta-}{O}$ bonds.

Isomerism of ethers. Three isomeric ethers $C_4H_{10}O$ exist:

$$C_2H_5OC_2H_5 \qquad CH_3OCH_2CH_2CH_3 \qquad CH_3OCH(CH_3)_2$$

Only the two alkyl groups in the molecules differ. Four alcohols are isomers of these ethers, p. 111; ethanol is an isomer of methoxymethane.

Preparation of ethers

1. *Dehydration of alcohols.* Concentrated sulphuric acid dehydrates excess ethanol at 410 K:

$$2C_2H_5OH(l) \rightarrow H_2O + \underset{\text{ethoxyethane}}{C_2H_5OC_2H_5(l)}$$

The acid raises the boiling point of the ethanol-acid mixture so that ethanol is not boiled off below 410 K, and it is a source of hydrogen ions or protons, which catalyse the dehydration. Some ethanol is protonated and then reacts with unchanged ethanol, a nucleophile:

$$C_2H_5OH(l) + H^+ \rightarrow \underset{\text{protonated ethanol}}{C_2H_5OH_2^+(l)}$$

$$C_2H_5OH(l) + C_2H_5OH_2^+(l) \rightarrow C_2H_5OC_2H_5(l) + H_3O^+$$

Methanol is dehydrated to methoxymethane in the same way. Both methanol and ethanol are dehydrated to ethers by passing their vapours under pressure over aluminium oxide (a dehydration catalyst) at 520 K.

Dehydration of alcohols is not a general method because higher alcohols dehydrate mainly to the alkenes.

In the dehydration of ethanol, one volume of acid can dehydrate about 50 volumes of alcohol, which is added continuously to a mixture of the ethanol and acid. The method is called the *continuous etherification process.* The acid does not last indefinitely because it becomes gradually diluted, part of it chars the ethanol and is reduced to sulphur dioxide, and some is used up in side-reactions producing esters.

2. *From sodium alkoxide or phenoxide and halogenoalkane* (*Williamson's synthesis*). The mixture is heated under reflux. A simple or a mixed ether can be obtained:

$$C_2H_5ONa + IC_2H_5 \rightarrow NaI(s) + C_2H_5OC_2H_5$$

$$C_6H_5ONa + ICH_3 \rightarrow NaI(s) + C_6H_5OCH_3$$

The sodium iodide is precipitated and the ether is obtained by fractional distillation. The method clearly shows the structure of the ether formed.

To prepare ethoxyethane, $C_2H_5OC_2H_5$ (diethyl ether)

(This experiment can be dangerous. It should not be done by pupils.)

Principle. Ethanol, in excess, is dehydrated at 410 K by concentrated sulphuric acid, which provides protons as catalyst:

$$C_2H_5OH + H^+ \rightarrow C_2H_5OH_2^+$$
protonated ethanol

$$C_2H_5OH_2^+ + HOC_2H_5 \rightarrow C_2H_5OC_2H_5 + H_2O + H^+$$

Mixing the reagents. Add ethanol (50 cm^3) to concentrated sulphuric acid (50 cm^3) in a 500 cm^3 distilling flask. Add the alcohol a few cm^3 at a time and shake; cool the flask from time to time to avoid loss of ethanol. Add broken porous pot or anti-bumping granules, which prevent bumping during distillation later on.

Distilling. Arrange the apparatus as in Fig. 10.1. A tap funnel containing more ethanol (100 cm^3) and a thermometer, reading to at least 470 K and dipping into the liquid, are in the neck of the flask. The receiver is a flask, joined to the condenser by an adaptor and surrounded by broken ice. A rubber tube connected to the side-tube of the adaptor has its open end down a sink so that ethoxy-ethane vapour cannot pass back over the bench to the source of heat and explode.

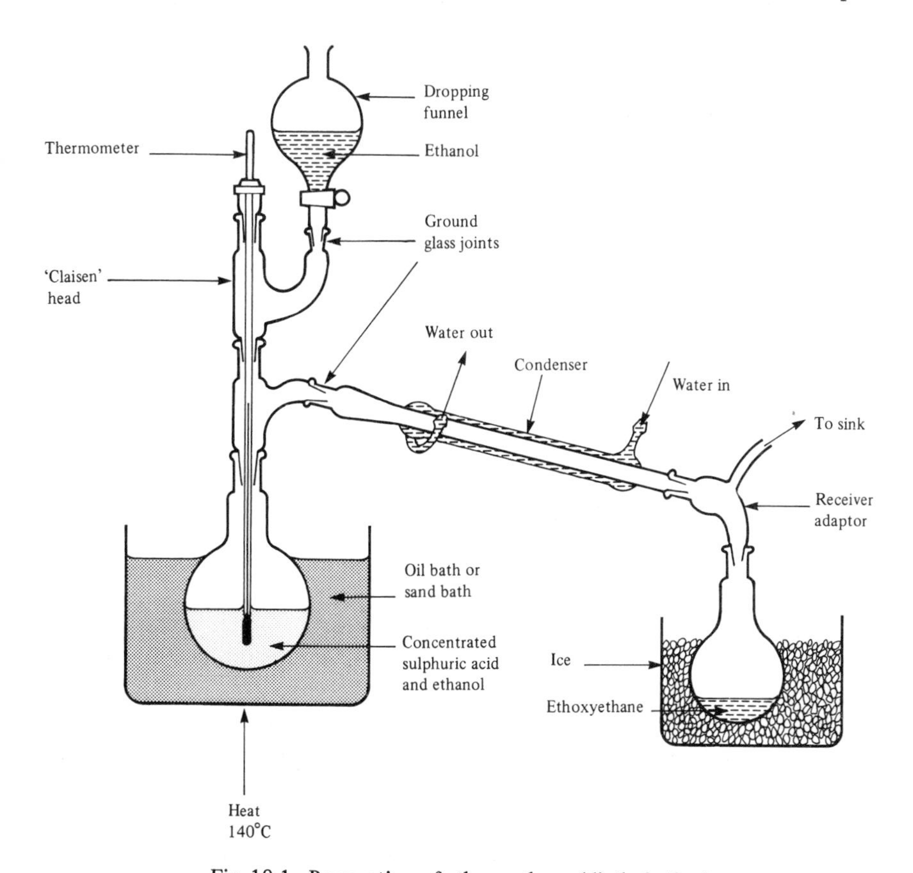

Fig. 10.1. Preparation of ethoxyethane (diethyl ether)

Heat the mixture in the distilling flask to 410 K, preferably by an electric hotplate (if a flame is used, a sand bath should be under the flask). Ethoxyethane distils over. Add ethanol from the tap funnel at the same rate as the ethoxyethane distils. Continue heating at 410 K for 5-10 minutes after all the ethanol has been added.

Purifying the ethoxyethane. Impurities include water and ethanol, and carbon dioxide and sulphur dioxide formed by the reactions:

$$C_2H_5OH + 2H_2SO_4 \rightarrow 2SO_2(g) + 5H_2O + 2C(s)$$

$$C + 2H_2SO_4 \rightarrow 2SO_2(g) + CO_2(g) + 2H_2O$$

(a) Pour the distillate into a separating funnel. Add dilute sodium hydroxide solution (about 20 cm^3) saturated with sodium chloride and shake. The alkali absorbs the acid gases. The sodium chloride reduces the solubility in water of the ethoxyethane. Let the lower aqueous layer run away.

(b) Transfer the moist ethoxyethane (the upper layer) to a flask. Add fused calcium chloride and either shake for 10 minutes or allow to stand for 24 hours. The chloride removes both water and ethanol.

(c) Pour the dry ethoxyethane into a small distilling flask. Distil in the usual way by placing a water bath containing boiling water around the flask (no flames should ever be near the apparatus). Collect the fraction passing over at 308 K, the boiling point. (Perfectly dry ethoxyethane can be obtained by drying the product with clean pieces of sodium.)

Properties of ethers

Physical properties. Ethoxyethane is a colourless volatile liquid with a sweet smell. It is an anaesthetic. Methoxymethane is a gas at ordinary temperatures (b.p. 248 K) and ethoxyethane boils at 307.7 K. Ethers are more volatile than the corresponding alcohols because they have no hydrogen bonds (p. 116). Ethers are only slightly soluble in water (contrast the lower alcohols). The density of ethoxyethane is 0.71 g cm^{-3} and therefore it floats on water. Usually ethoxyethane and water form two layers – the upper is a 1 per cent solution of water in ether and the lower is a 7 per cent solution of ether in water. Unlike alcohols, ethers contain no acidic hydrogen and the two lone pairs on the oxygen atoms make ethers basic substances. Therefore ethers are soluble in sulphuric acid (contrast the hydrocarbons). Ethoxyethane is a good solvent for covalent compounds, including many organic compounds, but not for salts. Ether extraction is used to separate organic compounds (e.g. phenol, benzaldehyde, phenylamine) from aqueous solution. The compounds are more soluble in ethoxyethane than in water and pass into the upper ether layer when the aqueous solution and ethoxyethane are shaken together. The ethoxyethane is easily separated from the compound by distillation because of its low boiling point.

Chemical properties. Ethers are much less reactive than alcohols and phenols because the oxygen is combined with two alkyl or aryl groups and not with an acidic hydrogen atom that can be substituted by another group. Ethers do not react with sodium, hardly react with cold phosphorus pentachloride, and have little or no reaction with most acids, and oxidizing agents. Because it is so unreactive, ethoxyethane is used as a solvent in several reactions such as those between halogenoalkanes and sodium (p. 98).

Ethoxyethane burns readily and it forms a dangerously explosive mixture with air:

$$(C_2H_5)_2O + 6O_2 \rightarrow 4CO_2 + 5H_2O$$

All flames must be extinguished when it is used, and it should be heated only on a water bath. On exposure to air and light, ethoxyethane gradually forms peroxides, which are non-volatile and explosive when heated strongly. Therefore never evaporate a solution to dryness using a flame or hot-plate, and only take ethoxyethane from a full bottle if you are going to distil it.

Excess warm concentrated sulphuric acid slowly converts ethoxyethane to ethyl hydrogensulphate

$$(C_2H_5)_2O(l) + 2H_2SO_4(l) \rightarrow 2C_2H_5HSO_4 + H_2O$$

Hot phosphorus pentachloride reacts slowly with ethoxyethane but forms no hydrogen chloride; the reaction is complex and C_2H_5Cl is formed.

Boiling an ether with hydriodic or hydrobromic acid causes fission:

$$(C_2H_5)_2O + 2HI \rightarrow 2C_2H_5I + H_2O$$

$$R'{-}O{-}R'' + 2HI \rightarrow R'I + R''I + H_2O$$

The fission takes place in two stages. An alcohol is the first product:

$$(C_2H_5)_2O + HI \rightarrow C_2H_5OH + C_2H_5I$$

The acid then converts the alcohol to the halogenoalkane. If a phenol is the first product, no further reaction occurs:

$$C_6H_5OCH_3 + HI \rightarrow C_6H_5OH + C_2H_5I$$

Hydriodic acid causes fission readily because its iodide ion is a powerful nucleophile.

Ethers are soluble in sulphuric acid and halogen acids because they are proton acceptors and therefore bases:

$$(CH_3)_2O + HCl \rightleftharpoons (CH_3)_2OH^+ + Cl^-$$

Compare: $H_2O + HCl \rightleftharpoons H_3O^+ + Cl^-$

Displayed (graphic) formula of ethoxyethane

Qualitative analysis shows that only carbon, hydrogen and oxygen are present. Quantitative analysis shows that the empirical formula is $C_4H_{10}O$. The relative density of the vapour is 37 and therefore the relative molecular mass of the ether is $2 \times 37 = 74$. The molecular formula is $C_4H_{10}O$ $(48 + 10 + 16 = 74)$.

Neither sodium nor cold phosphorus pentachloride reacts and therefore ethoxyethane does not contain a hydroxyl group.

Ethoxyethane can be prepared from iodoethane by the action of sodium ethoxide or dry silver oxide, and these reactions show that it contains two ethyl groups:

$$C_2H_5ONa + IC_2H_5 \rightarrow C_2H_5OC_2H_5 + NaI$$

$$2C_2H_5I + Ag_2O \rightarrow C_2H_5OC_2H_5 + 2AgI$$

The displayed formula is:

```
   H   H      H   H
   |   |      |   |
H—C—C—O—C—C—H
   |   |      |   |
   H   H      H   H
```

SUMMARY

Ethers

R—O—R′ (R and R′ = alkyl or aryl groups). Functional group: C—O—C. Polar bond is $\overset{\delta+}{C}—\overset{\delta-}{O}$.

Preparation

1. Dehydrate alcohols with conc H_2SO_4.

2. Reflux sodium alkoxide or phenoxide with halogenoalkane (Williamson's synthesis).

Properties of ethoxyethane, $C_2H_5OC_2H_5$

Unlike alcohols it has no action with Na, K, PCl_5 (cold), acids or oxidizing agents. Burns in air (explosive). Warm conc H_2SO_4 forms ethyl hydrogensulphate, $C_2H_5HSO_4$, and hot conc HI forms C_2H_5I. HBr and HCl react similarly but very slowly.

QUESTIONS

1. Write equations for the preparation from ethanol of ethoxyethane. Why is an excess of ethanol used in this preparation? What impurities are likely to be present in the product and how are they removed?

2. What is the action, if any, on ethoxyethane of (a) sodium, (b) phosphorus pentachloride, and (c) hydriodic acid? Deduce the structure of ethoxyethane from the reactions or lack of reactions.

3. Write displayed formulae for three isomers of molecular formula C_3H_8O. Outline chemical tests by which you could distinguish between them.

4. A compound ***X*** contains by mass 64.9 per cent of carbon, 13.5 per cent of hydrogen, and the rest is oxygen. ***X*** reacts with concentrated hydriodic acid to form compounds *Y* and *Z*; *Y* contains 89.44 per cent of iodine and *Z* contains 74.71 per cent of iodine. Identify the compounds ***X***, *Y* and *Z*.

5. Write the full structural or graphic formulae for all the ethers having the molecular formula $C_5H_{12}O$. How would you distinguish chemically between these isomeric ethers? Give equations, reagents and conditions for the reactions you mention. (C.)

6. Describe the reactions between each of the following pairs of compounds: ethanol and excess concentrated sulphuric acid, excess ethanol and concentrated sulphuric acid, ethene and bromine water, propene and hydrogen bromide. In each case give the type of reaction taking place, a balanced equation, and the probable mechanism of the change. (C.)

7. The composition by mass of two isomers is C = 60 per cent, O = 26.7 per cent, and H = 13.3 per cent. 0.60 g of each isomer occupied a volume of 0.336 dm^3 at 410 K and 760 mmHg pressure. Deduce their molecular formula. With sodium isomer ***A*** was unreactive but ***B*** produced an inflammable gas. ***B*** reacted with warm acidified potassium dichromate(VI) solution to form a product which, when purified, reacted with diamminesilver (Tollen's reagent) to form a precipitate of silver. Identify ***A*** and ***B*** and write simple equations for the two reactions of ***B*** mentioned above.

Chapter 11
CARBONYL COMPOUNDS

The carbonyl group, C=O, is present in aldehydes and ketones. In aldehydes, the group is joined to two hydrogen atoms in the first member of the series and to one hydrogen atom in the higher homologues. The functional group of aldehydes is therefore $-\underset{\substack{|\\ H}}{C}=O$, usually written $-CHO$. The general formula is RCHO (R = H, alkyl or aryl group). The first three aliphatic aldehydes are:

HCHO	CH_3CHO	C_2H_5CHO
$H_2C{=}O$	$H_3C{-}CH{=}O$	$CH_3CH_2{-}CH{=}O$
methanal (formaldehyde)	ethanal	propanal

In the systematic names, *-al* replaces the final *-e* of the name of the alkane with the same number of carbon atoms.

In ketones, the carbonyl group is joined to two alkyl or aryl groups, and the general formula is RCOR′ (R and R′ = an alkyl or aryl group).

CH_3COCH_3	$(CH_3)_2CHCOCH_3$	$CH_3CH_2COCH_2CH_3$
$H_3C{-}C({=}O){-}CH_3$	$(CH_3)_2CH{-}C({=}O){-}CH_3$	$CH_3CH_2{-}C({=}O){-}CH_2CH_3$
propanone (acetone)	3-methylbutanone	pentan-3-one

The suffix *-one* replaces the final *-e* of the corresponding alkane. The number indicates the carbon atom to which the oxygen atom is combined, i.e. the position of the carbonyl group. Since this group cannot be at the end of the chain in a ketone, names which end in 1-one are not correct. For example, $CH_3CH_2CH_2CH_2CHO$ is pentanal, an aldehyde, and not pentan-1-one.

Two compounds with aryl groups are:

C_6H_5CHO or C_6H_5–CHO (benzene ring)
benzaldehyde

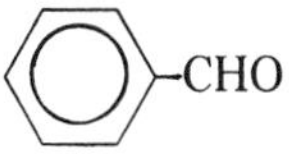

$C_6H_5COCH_3$ or C_6H_5–$COCH_3$ (benzene ring)
phenylethanone

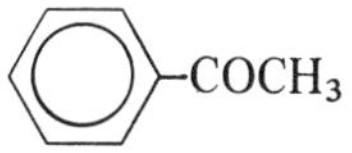

One cyclic ketone is cyclohexanone:

$C_6H_{10}O$ or (cyclohexane ring)=O

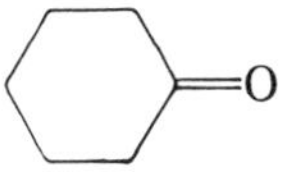

Isomerism exists among aldehydes and ketones. For example, C_3H_6O is the molecular formula of propanone and propanal, and C_4H_8O is the formula of butanone, butanal and 2-methylpropanal, $(CH_3)_2CHCHO$.

Preparation of aldehydes and ketones

1. *Oxidation of alcohols.* Heat an alcohol with sodium or potassium dichromate (VI) and sulphuric acid. Primary alcohols yield aldehydes and secondary alcohols yield ketones, p. 121.

$$CH_3CH_2OH \xrightarrow{Cr_2O_7^{2-}/H^+} CH_3CHO; \quad (CH_3)_2CHOH \xrightarrow{Cr_2O_7^{2-}/H^+} (CH_3)_2C{=}O$$

$$RCH_2OH \longrightarrow RCHO; \quad R'R''CH_2OH \longrightarrow R'R''CO$$

Dilute nitric acid or alkaline potassium manganate(VII) oxidize phenylmethanol:

$$C_6H_5CH_2OH \xrightarrow{\text{dil } HNO_3} C_6H_5CHO$$
benzaldehyde

Oxidation also occurs when the vapour of a primary or secondary alcohol, mixed with air or oxygen is passed over a hot copper catalyst:

$$2CH_3OH(g) + O_2 \rightarrow H_2O + HCHO(g)$$
methanal

Dehydrogenation occurs when alcohol vapour, without air, is passed over copper at about 600 K:

$$CH_3CH_2OH(g) \rightarrow H_2 + CH_3CHO; \quad (CH_3)_2CHOH(g) \rightarrow H_2 + (CH_3)_2CO$$

2. *From calcium or barium salts.* Powder the dry anhydrous salt of a carboxylic acid. Heat the powder strongly in a retort and condense the vapour evolved.

$$(HCOO)_2Ca \rightarrow CaCO_3 + H_2C{=}O$$

calcium methanoate (formate) → methanal

$$(CH_3COO)_2Ba \rightarrow BaCO_3 + (CH_3)_2C{=}O$$

barium ethanoate (acetate) → propanone

Mix calcium or barium methanoate with the salt of another carboxylic acid in order to obtain an aldehyde, e.g. with calcium ethanoate for ethanal and with calcium or barium benzoate for benzaldehyde:

$$Ca(OOCH)_2 + (CH_3COO)_2Ca \rightarrow 2CaCO_3 + 2CH_3CHO$$

$$Ba(OOCH)_2 + (C_6H_5COO)_2Ba \rightarrow 2BaCO_3 + 2C_6H_5CHO$$

3. *Hydrolysis of dihalogeno-compounds.* Heat with aqueous alkali:

$$CH_3CHBr_2 \xrightarrow{OH^-} CH_3CHO$$

1,1-dibromoethane

$$C_6H_5CHCl_2 \xrightarrow{OH^-} C_6H_5CHO$$

(dichloromethyl)benzene

4. *Miscellaneous methods.* Phenylethanone is produced in the following Friedel-Crafts reaction (p. 79):

$$C_6H_6(l) + ClCOCH_3(l) \xrightarrow{AlCl_3} HCl(g) + C_6H_5COCH_3(l)$$

Ethanal is formed by indirect hydration of ethyne (p. 64).

Manufacture of aldehydes and ketones

Methanal and ethanal are manufactured by passing the vapour of methanol and ethanol respectively, mixed with some air, over a silver catalyst at about 900 K. Oxidation of the alcohol by oxygen (an exothermic reaction) and dehydrogenation of the alcohol (an endothermic reaction) both occur. No external heat is required once the reaction has started:

$$CH_3OH + \tfrac{1}{2}O_2 \xrightarrow{Ag} HCHO + H_2O \qquad \Delta H = -154\,kJ$$

$$CH_3OH \xrightarrow{Ag} HCHO + H_2 \qquad \Delta H = +120\,kJ$$

Propanone is manufactured by dehydrogenation of propan-2-ol, using hot copper as catalyst. Propanone is present in wood spirit (p. 119).

Ethanal is obtained by oxidation of ethene with air or oxygen, using special catalysts:

$$2CH_2{=}CH_2 + O_2 \rightarrow 2CH_3CHO$$

Benzaldehyde is manufactured by hydrolysis of (dichloromethyl)benzene, $C_6H_5CHCl_2$, using sodium carbonate or calcium hydroxide. The product always contains chlorine as impurity. The purer compound, required to make perfumes and flavours, is made by oxidizing methylbenzene vapour with air in the presence of a catalyst:

$$C_6H_5CH_3 + O_2 \rightarrow H_2O + C_6H_5CHO$$

To prepare ethanal

Principle. Ethanol is oxidized by sodium dichromate(VI) and sulphuric acid:

$$CH_3CH_2OH \xrightarrow[H_2SO_4]{Na_2Cr_2O_7} CH_3CHO$$

Since ethanal is readily oxidized to ethanoic acid, excess oxidizing agent must be avoided and the aldehyde must leave the reaction mixture as soon as it is formed. This is done by adding a mixture of ethanol and acid slowly to the dichromate(VI) solution. Sodium dichromate is more soluble than potassium dichromate in aqueous ethanol and therefore the sodium salt is preferred.

Method. Dissolve sodium dichromate(VI) (6 g) in about 18 cm^3 of water. Place the solution in a distillation flask (50 cm^3), Fig. 11.1.

Mix ethanol (7 cm^3) with concentrated sulphuric acid (4 cm^3) by adding the alcohol slowly and carefully to the acid. Place this mixture in the tap funnel.

Heat the dichromate solution to boiling. Turn out the flame and add the alcohol-acid mixture from the tap funnel, drop by drop. Heat is evolved and the first reaction is vigorous. When the mixture stops boiling, heat gently. Collect about 6 cm^3 of distillate in the cooled conical flask. It contains ethanal, water, and some unchanged ethanol.

Purification. Distil the ethanal (b.p. 294 K) from the mixture and collect it in ether. Pass dry ammonia into the solution. Crystals of an addition compound, ethanal ammonia, are formed. Filter off the crystals, distil with dilute sulphuric acid, and collect pure ethanal in a receiver cooled by ice.

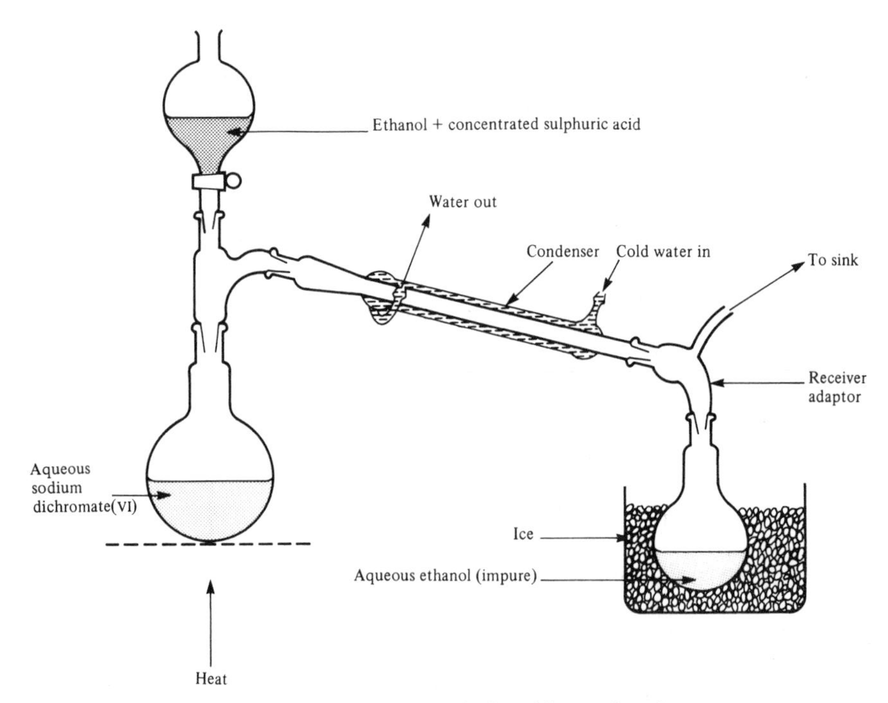

Fig. 11.1. Preparation of ethanal from ethanol

Tests on aldehydes and ketones

In the following tests use formalin (a 40 per cent solution of methanal in water), aqueous ethanal, benzaldehyde, propanone, phenylethanone, and cyclohexanone.

1. Observe the colour and smell of each compound.

2. Add a few cm^3 of formalin to a watch-glass and evaporate to dryness on a water bath. Observe the residue.

Place some of the residue on a crucible lid and warm. Note any change and identify the vapour by its smell. (Do not repeat test 2 with the other compounds.)

3. *Sodium hydroxide.* Warm aqueous ethanal with 20 per cent (5 M) sodium hydroxide solution. Observe the colour changes and the product. Repeat this test with formalin and propanone only.

Warm 1 cm^3 of benzaldehyde with 2 cm^3 of sodium hydroxide, about 5 M. Shake well during the warming. Allow to cool and add concentrated hydrochloric acid carefully.

4. *Air.* Place a few drops of benzaldehyde on a watch-glass and leave in the air for a day or so. Note any change.

5. *Diamminesilver*, $[Ag(NH_3)_2]^+$, *Tollen's reagent* (Ag^+ as oxidizing agent). To 2 cm^3 of aqueous silver nitrate in a *clean* test-tube, add ammonia solution until the precipitate first formed just dissolves. The reagent contains $[Ag(NH_3)_2]^+$.

Dilute formalin about 20 times and add a few drops of the diluted solution to Tollen's reagent. Place the tube in a beaker of warm water. Observe any change.

Repeat with ethanal, benzaldehyde and propanone.

(Pour away any unused ammoniacal solution and also the liquid in the tubes. An explosive compound may form if these solutions evaporate to dryness.)

6. *Fehling's solution* (p. 167) (Cu^{2+} as oxidizing agent). Fehling's solution contains copper sulphate, sodium hydroxide and sodium potassium 2,3-dihydroxybutanedioate (sodium potassium tartrate). It contains copper as part of a complex ion but may be regarded as Cu^{2+} in alkaline solution.

Add a few drops of formalin to the solution and warm.

Repeat the test with ethanal, benzaldehyde and propanone.

7. *Potassium manganate(VII).* Add cold, very dilute, potassium manganate(VII), acidified with sulphuric acid, to formalin. Observe any colour change.

Repeat the test with ethanal, propanone and phenylethanone. Warm if necessary.

Use acidified potassium dichromate(VI) instead of the potassium manganate(VII) with ethanal and propanone.

8. *Sodium hydrogensulphite.* Freshly prepared solution must be used. Dissolve the hydrogensulphite in water and saturate with sulphur dioxide. A pale-green solution forms.

Add 4 cm^3 of the hydrogensulphite solution to each of four tubes. To the tubes add about 2 cm^3 of (a) concentrated ethanal solution, (b) benzaldehyde, (c) propanone, and (d) cyclohexanone. Some changes occur in a few minutes, but others take place only after 1 or 2 hours.

9. *2,4-dinitrophenylhydrazine.* Add 4 cm^3 of concentrated sulphuric acid to 2 g of 2,4-DNP, and then carefully add 30 cm^3 of methanol. Warm gently until a solution forms, and add 10 cm^3 of distilled water.

Add 5 cm^3 of the above solution to each of several tubes. Add various aldehydes and ketones separately to the tubes. Observe any reactions.

10. *Semicarbazide.* Dissolve 2 g of semicarbazidium chloride in 20 cm^3 of water containing 3 g of sodium ethanoate crystals. Divide the solution into two tubes. To one add 1 cm^3 of propanone and to the other 1 cm^3 of cyclohexanone. Cork the tubes, shake well, and allow to stand.

11. *Hydroxylamine.* Dissolve 1 g of hydroxylammonium chloride in 4 cm^3 of warm water containing 1.5 g of sodium ethanoate crystals. Add 1 cm^3 of cyclohexanone. Cork the tube, shake well, and allow to stand.

Physical properties of aldehydes and ketones

Methanal is a colourless gas (the only gaseous aldehyde) with an irritating smell. It is readily soluble in water, probably because it forms a hydrate (see later). Its aqueous solution, containing 40 per cent methanal by mass, is called formalin. Methanal is poisonous. It hardens and preserves biological tissues, probably because it coagulates the proteins in them.

Ethanal is a colourless volatile liquid with a pungent smell that is not unpleasant. It is completely miscible with water.

Propanone is a colourless liquid with a pleasant characteristic smell. It is completely miscible with water, alcohols and ether, and is a good solvent for many organic compounds, including ethyne.

Benzaldehyde is a colourless liquid which gradually turns yellow on keeping. It has a strong pleasant smell of almonds and is poisonous. It is sparingly soluble in and denser than water. It is completely miscible with ethanol and ether.

Phenylethanone is a solid (m.p. 293 K) with a smell of almonds.

The boiling points of aldehydes and ketones are given in the following tables, together with the melting points of the derivatives formed with 2,4-dinitrophenylhydrazine. These derivatives are formed readily, are stable, crystalline, and can be readily recrystallized from an ethanol-water mixture (1 : 1). They are used to identify aldehydes, ketones and some sugars.

Name	*Formula*	*B.p./K*	*M.p./K 2,4-DNP derivative*
Methanal	$HCHO$	254	440
Ethanal	CH_3CHO	294	437
Propanal	CH_3CH_2CHO	321	429
Butanal	$CH_3CH_2CH_2CHO$	348	396
2-Methylpropanal	$(CH_3)_2CHCHO$	337	460
Benzaldehyde	C_6H_5CHO	451	510
Propanone	CH_3COCH_3	329	401
Butanone	$CH_3CH_2COCH_3$	353	388
Pentan-2-one	$CH_3CH_2CH_2COCH_3$	375	414
Pentan-3-one	$CH_3CH_2COCH_2CH_3$	375	429
3-Methylbutanone	$(CH_3)_2CHCOCH_3$	367	390
Cyclohexanone	$C_6H_{10}O$	429	435
Phenylethanone	$C_6H_5COCH_3$	475	523

The carbonyl group in aldehydes and ketones

The double bond of the carbonyl group consists of a σ- and a π-bond, and is strongly polarized because oxygen is much more electronegative than carbon (p. 93):

$$>\overset{\delta+}{C}=\overset{\delta-}{O}$$

Thus the carbon atom bears a partial positive charge and the oxygen atom bears a partial negative charge, and on approach of a nucleophile the π-electrons, which are more easily displaced than the σ-electrons, move towards the oxygen atom. In many reactions, nucleophiles such as NH_3, CN^- and $HSO_3{}^-$ attack the carbon atom, and the oxygen atom then acquires an electrophile, usually a proton:

$$Nu{:}^- + >C=\ddot{O}{:} \rightarrow Nu-\overset{|}{\underset{|}{C}}-\ddot{O}{:}^- \quad \text{then} \quad Nu-\overset{|}{\underset{|}{C}}-O^- + H^+ \rightarrow Nu-\overset{|}{\underset{|}{C}}-OH$$

Addition reactions are therefore characteristic of aldehydes and ketones. Sometimes the saturated addition product is unstable and elimination of water occurs, and the final product is unsaturated.

Addition to an alkene double bond is quite different (p. 49). C=C is not polarized, and here the π-electron cloud of the double bond reacts with an electrophile in the first stage of addition.

The general order of reactivity of aldehydes and ketones is:

$$HCHO > RCHO > ArCHO > R_2CO > Ar_2CO \quad (R = \text{alkyl},\ Ar = \text{aryl})$$

Methanal is most reactive. The alkyl group R in RCHO tends to donate or release electrons (p. 189) to the carbon of the carbonyl group by induction, and make it less susceptible to attack by nucleophiles. The aromatic ring Ar in ArCHO releases electrons to the carbonyl group, and therefore C_6H_5CHO is less reactive than CH_3CHO. Ketones, which have two alkyl or aryl groups, are less reactive than aldehydes. Since large alkyl or aryl groups hinder the approach of nucleophiles more than small groups, they also decrease the reactivity of the compounds. Therefore, $CH_3CH_2CH_2CHO$ is less reactive than CH_3CHO, and $C_3H_7COC_2H_5$ is less reactive than CH_3COCH_3. (In some reactions benzaldehyde is less reactive than ethanal merely because it is much less soluble in water.)

Chemical properties of aldehydes and ketones

1. *Polymerization.* Ketones do not polymerize. Methanal polymerizes most readily and its aqueous solution forms a white solid polymer when evaporated:

$$n\text{HCHO(aq)} \xrightarrow{\text{heat}} \underset{\text{poly(methanal)}}{(\text{HCHO})_n(\text{s})} \quad (n = \text{about } 40)$$

Gaseous methanal is evolved when this solid is heated.

The trimer $(HCHO)_3$ is formed when methanal vapour is cooled. It is a white solid, soluble in water. The trimer is readily formed by traces of acids and alkalis.

Calcium hydroxide solution slowly turns methanal to a mixture of sugars:

$$6\text{HCHO(aq)} \xrightarrow{\text{Ca(OH)}_2} \underset{\text{formose}}{\text{C}_6\text{H}_{12}\text{O}_6\text{(aq)}}$$

Very dilute sodium hydroxide or carbonate converts ethanal to a colourless liquid called 3-hydroxybutanal (aldol). The *aldol reaction* is an addition reaction and aldol is a dimer:

$$\text{CH}_3\text{CHO(aq)} + \text{HCH}_2\text{CHO(aq)} \rightarrow \underset{\text{3-hydroxybutanal}}{\text{CH}_3\text{CHOHCH}_2\text{CHO(aq)}}$$

The product is an *ald*ehyde and an alcoh*ol*, hence the name aldol. Propanone also forms its dimer under special conditions:

$$(\text{CH}_3)_2\text{CO} + \text{HCH}_2\text{COCH}_3 \rightarrow (\text{CH}_3)_2\text{COHCH}_2\text{COCH}_3$$

One drop of cold concentrated sulphuric acid converts ethanal to ethanal trimer, $(CH_3CHO)_3$, a colourless liquid. Ethanal tetramer, $(CH_3CHO)_4$, a solid, is formed by the action of mineral acids below 273 K. Warm concentrated sodium hydroxide polymerizes ethanal to a brown resin of unknown structure.

Methanal and benzaldehyde do not form resins when warmed with sodium (or potassium) hydroxide. Instead, oxidation to an acid and reduction to an alcohol occur simultaneously in a *Cannizzaro reaction*:

$$2HCHO \xrightarrow[\text{or KOH}]{\text{NaOH}} CH_3OH + HCOOH$$

$$2C_6H_5CHO \longrightarrow C_6H_5COOH + C_6H_5CH_2OH$$

2. *Reducing reactions.* Ketones are weak reducing agents, and mild oxidizing agents have no action on them. Aldehydes are reducing agents because a hydrogen atom can be removed without breaking the carbon–carbon bonds. Aldehydes are oxidized to acids, but methanoic acid may be further oxidized to carbon dioxide:

$$HCHO \xrightarrow{[O]} HCOOH \xrightarrow{[O]} CO_2 + H_2O$$

$$CH_3CHO \longrightarrow CH_3COOH$$

$$CH_3COCH_3 \longrightarrow CH_3COOH + CO_2 + H_2O$$

Vigorous oxidizing agents such as sodium (or potassium) dichromate(VI) and sulphuric acid oxidize ketones slowly to a carboxylic acid and carbon dioxide.

(a) Benzaldehyde oxidizes readily in air, and bottles of the aldehyde often contain white crystals of benzoic acid.

(b) *Tollen's reagent* (*silver mirror test*). This reagent uses Ag^+ as the oxidizing agent. The ion is present as the complex diamminesilver(I) ion $[Ag(NH_3)_2]^+$. Aldehydes reduce the reagent to silver, which is formed as a black precipitate, or, in a clean glass tube, as a silver mirror. Benzaldehyde reacts slowly, but ketones do not react.

$$CH_3CHO(l) + 2Ag^+(aq) + 3OH^-(aq) \rightarrow 2Ag(s) + CH_3COO^-(aq) + 2H_2O$$

or $CH_3CHO + 2Ag[(NH_3)_2]^+ + 2OH^- \rightarrow 2Ag(s) + CH_3COONH_4 + 3NH_3 + H_2O$

(c) *Fehling's solution.* Cu^{2+} is the oxidizing agent and it is present as a complex ion, in alkaline solution. Aldehydes form a reddish brown precipitate of copper(I) oxide, but benzaldehyde and ketones have no action:

$$RCHO(l) + 2Cu^{2+}(aq) + 5OH^-(aq) \rightarrow Cu_2O(s) + RCOO^-(aq) + 3H_2O$$

or $RCHO + 2CuO \rightarrow Cu_2O(s) + RCOOH$

(d) *Potassium manganate(VII) and dichromate(VI).* Aldehydes and ketones decolorize purple potassium manganate(VII), acidified with dilute sulphuric acid; colourless manganese(II) ions are formed:

$$CH_3CHO \xrightarrow{MnO_4^-/H^+} CH_3COOH\ (+\ Mn^{2+}, \text{colourless})$$

Aldehydes and ketones are oxidized by sodium (or potassium) dichromate(VI), acidified with dilute sulphuric acid; the orange solution turns green:

$$CH_3COCH_3 \xrightarrow{Cr_2O_7^{2-}/H^+} CH_3COOH + CO_2 \ (+\ Cr^{3+}, \text{green})$$

3. *Addition reactions of aldehydes and ketones.* Carbonyl compounds are unsaturated because of the double bond C=O. A molecule HX (X = univalent group) can therefore add to a molecule of an aldehyde or ketone to form one molecule of a saturated compound:

$$>C{=}O + HX \rightarrow\ >C(OH)X$$

During the addition reaction, a nucleophile attacks the positively charged carbon of the carbonyl group. Some nucleophiles that do this are H_2O:, :NH_3, CN^- and HSO_3^-. Water and ammonia molecules have at least one lone pair of electrons (denoted by ':') and the other reagents are anions.

(a) *Water.* Methanal in aqueous solution is hydrated:

$$H{-}\underset{}{\overset{H}{\overset{|}{C}}}{=}O + HOH \rightleftharpoons H{-}\underset{\underset{OH}{|}}{\overset{\overset{H}{|}}{C}}{-}OH$$

dihydroxymethane

Other aldehydes and ketones are slightly hydrated.

(b) *Hydrogen.* Aldehydes are reduced to primary alcohols and ketones to secondary alcohols. Reducing agents are hydrogen with a nickel catalyst, sodium amalgam and water, lithium tetrahydridoaluminate(III), Li[AlH_4], and sodium tetrahydridoborate(III), Na[BH_4].

$$RCHO \rightarrow RCH_2OH; \quad RCOR' \rightarrow RCHOHR'$$

Lithium tetrahydridoaluminate(III) is a greyish-white solid, and its solution in ether is used because water decomposes it. Sodium tetrahydridoborate (III) is safer to use because it is soluble in a mixture of water and methanol. The complex ions $[AlH_4]^-$ and $[BH_4]^-$ supply hydride ions H:$^-$, which are powerful nucleophiles. (These complex ions do not reduce alkenes to alkanes.)

The reduction of C=O and C=C by gaseous hydrogen in the presence of nickel is the only addition reaction common to aldehydes, ketones and alkenes.

(c) *Ammonia.* White ethanal-ammonia is precipitated when ammonia is passed into an ethereal solution of ethanal:

$$CH_3\overset{\overset{H}{|}}{C}=O + NH_3 \rightarrow CH_3\overset{\overset{H}{|}}{\underset{\underset{NH_2}{|}}{C}}-OH$$

ethanal-ammonia

Other aliphatic aldehydes, except methanal, form similar addition compounds. Methanal, benzaldehyde and ketones may form unstable addition compounds at first but secondary reactions at once occur, usually involving loss of water. The final products are very complex compounds.

(d) *Hydrogen cyanide.* Aldehydes and ketones combine with hydrogen cyanide which contains a little alkali (to provide CN^- ions). The product, called a *hydroxynitrile* (cyanohydrin) contains a cyanide group and a hydroxyl group joined to the same carbon atom. The nucleophile CN^- adds first:

$$\underset{/\backslash}{\overset{O}{\overset{\|}{C}}} + CN^- \rightarrow -\overset{\overset{O^-}{|}}{\underset{|}{C}}-CN \xrightarrow{H^+} -\overset{\overset{OH}{|}}{\underset{|}{C}}-CN; \quad \overset{R}{\underset{H}{>}}C=O + HCN \rightarrow \overset{R}{\underset{H}{>}}C\overset{OH}{\underset{CN}{<}}$$

The addition products are readily hydrolysed by dilute acid to hydroxycarboxylic acids:

$$CH_3\overset{\overset{H}{|}}{C}=O \xrightarrow{HCN} CH_3\overset{\overset{H}{|}}{\underset{\underset{CN}{|}}{C}}-OH \xrightarrow{hydrolysis} CH_3\overset{\overset{H}{|}}{\underset{\underset{COOH}{|}}{C}}-OH$$

2-hydroxypropanenitrile 2-hydroxypropanoic acid

2-hydroxy-2-methylpropanenitrile $(CH_3)_2COHCN$ (from propanone) yields the hydroxyacid $(CH_3)_2COHCOOH$. The general reactions with aldehydes are:

$$RCHO \xrightarrow{HCN} RCHOHCN \xrightarrow{hydrolysis} RCHOHCOOH$$

(Benzaldehyde does not react with hydrogen cyanide.)

(e) *Sodium hydrogensulphite.* Aldehydes and methyl ketones (CH_3COR) combine with a cold concentrated solution of the hydrogensulphite. The addition compounds are white solids:

$$RC(H){=}O(l) + NaHSO_3(aq) \rightarrow R-C(H)(SO_3Na)-OH(s)$$

$$CH_3C(R){=}O(l) + NaHSO_3(aq) \rightarrow CH_3-C(R)(SO_3Na)-OH(s)$$

ethanal hydrogensulphite

Dilute sulphuric acid liberates the free aldehyde or methyl ketone from the addition compounds. Therefore they are used to separate carbonyl compounds from other organic compounds which do not react with sodium hydrogensulphite.

Ketones such as $(C_2H_5)_2CO$ which have no methyl group do not form hydrogensulphite addition compounds.

4. *Addition-elimination reactions of aldehydes and ketones.* Hydroxylamine, NH_2OH, and phenylhydrazine, $C_6H_5NHNH_2$, can be regarded as derivatives of ammonia in which OH and C_6H_5NH respectively replace one hydrogen atom of an ammonia molecule. Both compounds add to aldehydes and ketones and then the addition products lose water; addition is followed by elimination of the elements of water. Addition-elimination reactions are sometimes called *condensation reactions.*

The general equation for the addition-elimination reactions of aldehydes and ketones with ammonia derivatives H_2NX is:

$$>C{=}O + H_2NX \rightarrow >C(NHX)-OH \xrightarrow{-H_2O} >C{=}NX$$

(a) *Hydroxylamine. Oximes* are formed:

$$(CH_3)(H)C{=}O + H_2NOH \xrightarrow{-H_2O} (CH_3)(H)C{=}NOH$$

ethanal oxime

Oximes can be reduced to amines by sodium and ethanol or by hydrogen in the presence of nickel:

$$(CH_3)_2C{=}O \xrightarrow{NH_2OH} (CH_3)_2C{=}NOH \xrightarrow{H_2} (CH_3)_2CHNH_2$$

propanone oxime — 1-methylethylamine

(b) *2,4-dinitrophenylhydrazine.* Aldehydes and ketones form *hydrazones* with hydrazine, NH_2NH_2, *phenylhydrazones* with phenylhydrazine, $C_6H_6NHNH_2$, and *2,4-dinitrophenylhydrazones* with 2,4-dinitrophenylhydrazine, $C_6H_3(NO_2)_2NHNH_2$ or

$$NO_2-C_6H_3(NO_2)-NHNH_2$$

The products are used to identify aldehydes, ketones and some sugars because they are yellow, orange or red solids, easily purified, and have sharp melting points. The 2,4-dinitro products are most suitable because their melting points are highest and in a convenient range.

$$>C{=}O + H_2NNHC_6H_5 \xrightarrow{-H_2O} >C{=}NNHC_6H_5$$

$$>C{=}O + H_2NNHC_6H_3(NO_2)_2 \xrightarrow{-H_2O} >C{=}NNHC_6H_3(NO_2)_2$$

$$(CH_3)_2C{=}O + NH_2-NH-C_6H_3(O_2N)-NO_2 \xrightarrow{-H_2O} (CH_3)_2C{=}N-NH-C_6H_3(O_2N)-NO_2$$

(In reaction 10, p. 150, semicarbazide, $H_2NNHCONH_2$, was used to form a solid semicarbazone, $(CH_3)_2C{=}NNHCONH_2$, which is more easily prepared than a solid oxime.)

(c) *Alcohols* (*hemiacetal and acetal formation*). An aldehyde and an alcohol, in the presence of dry hydrogen chloride as a catalyst, condense to form an acetal:

$$CH_3CHO + 2HOCH_3 \rightleftharpoons H_2O + CH_3CH(OCH_3)_2$$

1,1-dimethoxyethane
(dimethyl acetal)

$$CH_3CHO + 2HOC_2H_5 \rightleftharpoons H_2O + CH_3CH(OC_2H_5)_2$$

1,1-diethoxyethane
(diethyl acetal)

Each reaction takes place in two stages. The intermediate product is a hemiacetal (half acetal) because it is only half converted to the final acetal. The general equation is:

$$\underset{\text{aldehyde}}{R-\overset{\displaystyle H}{\overset{|}{C}}=O} \xrightleftharpoons{R'OH} \underset{\text{hemiacetal}}{R-\overset{\displaystyle H}{\overset{|}{\underset{\underset{\displaystyle OR'}{|}}{C}}}-OH} \xrightleftharpoons{R'OH} \underset{\text{acetal}}{R-\overset{\displaystyle H}{\overset{|}{\underset{\underset{\displaystyle OR'}{|}}{C}}}-OR'}$$

Aqueous acids liberate the original aldehydes from acetals. With other reagents, including bases, acetals are stable.

Ketones do not form ketals in the same way.

5. *Phosphorus pentachloride.* This reagent converts the C=O group to CCl_2 and no hydrogen chloride is evolved (contrast alcohols):

$$CH_3CHO(l) + PCl_5(s) \rightarrow PCl_3O + \underset{\text{1,1-dichloroethane}}{CH_3CHCl_2(l)}$$

$$CH_3COCH_3(l) + PCl_5(s) \rightarrow PCl_3O + \underset{\text{2,2-dichloropropane}}{CH_3CCl_2CH_3(l)}$$

Benzaldehyde forms $C_6H_5CHCl_2$, (dichloromethyl)benzene.

6. *Chlorine.* Ethanal is converted to trichloroethanal:

$$CH_3CHO + 3Cl_2 \rightarrow CCl_3CHO + 3HCl$$

Trichloroethanal and water form a stable solid addition product which is unusual in having two hydroxyl groups joined to one carbon atom, $CCl_3CH(OH)_2$ (2,2,2-trichloroethanediol).

7. *Trihalomethane (haloform) reaction.* Methyl ketones, ethanal (and alcohols with the CH_3CHOH- group) form a yellow precipitate of triiodomethane, CHI_3, with iodine and alkali (p. 112).

$$CH_3\overset{\displaystyle R}{\overset{|}{C}}=O \xrightarrow{I_2} CI_3\overset{\displaystyle R}{\overset{|}{C}}=O \xrightarrow{\text{alkali}} CHI_3 + RCOOH$$

Trichloromethane and tribromomethane are formed by similar reactions.

8. *Ring reactions of benzaldehyde.* Chlorination, nitration and sulphonation of benzaldehyde are possible. Chlorine in the presence of a halogen-carrier yields 3-chlorobenzaldehyde, a mixture of concentrated nitric and sulphuric acids forms 3-nitrobenzaldehyde, and sulphuric acid alone forms the 3-sulphonic acid derivative.

Resemblances between aldehydes and ketones

Ethanal and propanone are colourless liquids which are completely miscible with water but do not form stable hydrates. They are prepared by oxidation or dehydrogenation of alcohols and by the action of heat on calcium or barium salts of carboxylic acids. They are formed by hydrolysis of dichloro-compounds, CH_3CHBr_2, CH_3CHCl_2 and $CH_3CCl_2CH_3$.

Most reactions of aldehydes and ketones are similar. Ethanal and propanone both:

(a) *polymerize* to the dimers $(C_2H_4O)_2$ and $(C_3H_6O)_2$;

(b) are *oxidized* to acids by acidified potassium manganate(VII);

(c) form *addition products*:

alcohols with reducing agents,
hydroxynitriles with hydrogen cyanide,
hydrogensulphite compounds with sodium hydrogensulphite and
hemiacetals and hemiketals with alcohols;

(d) form *addition-elimination* (*condensation*) *products*:

oximes with hydroxylamine, NH_2OH,
hydrazones with hydrazine, NH_2NH_2,
phenylhydrazones with phenylhydrazine, $C_6H_5NHNH_2$,
2,4-dinitrophenylhydrazones with 2,4-dinitrophenylhydrazine,
$C_6H_3(NO_2)_2NHNH_2$;

(e) form *dichloro-derivatives* (and no hydrogen chloride) with phosphorus pentachloride;

(f) form *triiodomethane* with iodine and alkali.

Differences between aldehydes and ketones

Polymerization of aldehydes occurs readily, but ketones do not polymerize, although propanone forms its dimer, $(C_3H_6O)_2$, under special conditions. Ethanal readily forms its dimer, $CH_3CHOHCH_2CHO$, its trimer and tetramer, $(C_2H_4O)_3$ and $(C_2H_4O)_4$, and a brown resin.

Oxidation by Tollen's reagent and Fehling's solution occurs with aldehydes but not with ketones. Aldehydes are oxidized to acids much more readily than ketones are because the hydrogen atom of their CHO group can be converted to OH whereas ketones can be oxidized only by breaking a C—C bond.

Oxidation by potassium manganate(VII) or dichromate(VI) converts aldehydes to acids and ketones to acids and carbon dioxide, that is, the acid molecule contains fewer carbon atoms than the ketone molecule:

$$CH_3CHO \xrightarrow{[O]} CH_3COOH; \qquad (CH_3)_2CO \xrightarrow{4[O]} CH_3COOH + CO_2 + H_2O$$

Reduction of aldehydes produces primary alcohols whereas ketones produce secondary alcohols.

Ammonia forms addition compounds with aldehydes (except methanal) but not with ketones.

Differences between aldehydes

Methanal differs from all other aldehydes in having two hydrogen atoms attached to the carbonyl group. It is the strongest reducing agent of the aldehydes, and will even reduce mercury(II) chloride solution to a white precipitate of mercury(I) chloride and then to a grey precipitate of mercury:

$$2HgCl_2(aq) + HCHO(aq) + H_2O \rightarrow Hg_2Cl_2(s) + HCOOH(aq) + 2HCl$$

$$Hg_2Cl_2(s) + HCHO(aq) + H_2O \rightarrow 2Hg(l) + HCOOH(aq) + 2HCl$$

Hydration of methanal, forming $CH_2(OH)_2$, occurs to a much greater extent than hydration of other aldehydes. Methanal hydrate polymerizes readily to white solid poly(methanal).

Ammonia forms a complex addition compound with methanal but it forms addition compounds, $RCHOHNH_2$, with other aldehydes.

Differences between benzaldehyde and ethanal

Benzaldehyde has a characteristic smell of almonds, is only sparingly soluble in water, and has a much higher boiling point than ethanal. It does not form an addition compound with ammonia. Warm sodium hydroxide converts ethanal to a brown resin, but benzaldehyde forms phenylmethanol and benzoic acid by the Cannizzaro reaction. Benzaldehyde is a weaker reducing agent: it does not reduce Fehling's solution (although it is oxidized slowly by Tollen's reagent and by air). Its aromatic ring can be chlorinated, nitrated and sulphonated, and these reactions are not possible with ethanal.

Uses of aldehydes and ketones

Methanal is used to preserve anatomical and biological specimens, to sterilize surgical instruments, and to disinfect rooms and clothing contaminated with disease germs. It is used together with phenol to form Bakelite (p. 133) and with urea to form another plastic, used in glues, lacquers and cements. A strong polymer has been manufactured from the gaseous aldehyde and is used to replace metals in some light machinery. The polymer is a polyoxymethylene: $\left[O-CH_2-O-CH_2-O-CH_2\right]_n$.

Propanone is used as a solvent for acetylene in storage cylinders, for various lacquers and perfumes, for the explosives used to make cordite, and for substances used in the making of photographic film. Trichloromethane and triiodomethane are manufactured from propanone.

Benzaldehyde is used in some flavours and perfumes and to prepare certain dyes. 2,2,2-trichloroethanediol is a soporific.

SUMMARY

Aldehydes and ketones

Both contain the carbonyl group $>C=O$. Aldehydes RCHO or $\begin{matrix} R \\ H \end{matrix}\!\!>\!C=O$.

Ketones RCOR′ or $\begin{matrix} R \\ R' \end{matrix}\!\!>\!C=O$.

Preparation
Ethanal, CH_3CHO:

Oxidize primary alcohol with $Cr_2O_7^{2-}/H^+$.
Heat calcium (or barium) ethanoate and methanoate.

Propanone, CH_3COCH_3:

Oxidize secondary alcohol. Heat calcium (or barium) ethanoate.
Heat calcium (or barium) ethanoate.

Phenylethanone, $C_6H_5COCH_3$, is obtained by the Friedel-Crafts reaction:

$$C_6H_6(l) + CH_3COCl/AlCl_3$$

Properties
The double bond in C=O consists of a σ- and a π-bond, and is readily polarized. Nucleophiles attack the positive C atom. Aldehydes are more reactive.

Comparison of properties of ethanal and benzaldehyde

Reagent/reaction/ property	*Ethanal, CH_3CHO*	*Benzaldehyde, C_6H_5CHO*
Smell	Pungent	Almonds
Water	Soluble	Sparingly soluble
Boiling point	Lower	Much higher
Differences		
NaOH(aq)	Dimer (aldol) or brown resin	Cannizzaro reaction. $C_6H_5COONa + C_6H_5CH_2OH$
One drop H_2SO_4	Trimer, paraldehyde	No similar reaction
Tollen's reagent	Silver mirror	No reaction (or very slow)
Fehling's solution	Red copper(I) oxide	No reaction
NH_3(ether)	$CH_3CHOHNH_2$	Complex compound
I_2 + alkali	CHI_3, yellow precipitate	No reaction
$Cl_2 + AlCl_3$	Similar reactions are not possible.	3-ClC_6H_4CHO chlorination
$HNO_3 + H_2SO_4$ (conc)		3-$NO_2C_6H_4CHO$ nitration
H_2SO_4 (conc), hot		3-$SO_2OHC_6H_4CHO$ sulphonation
Similarities		
MnO_4^-/H^+, oxidation	CH_3COOH ethanoic acid	C_6H_5COOH benzoic acid
H_2/Ni, Li[AlH_4] (ether), reduction	CH_3CH_2OH ethanol	$C_6H_5CH_2OH$ phenylmethanol
HCN forms hydroxynitriles	$CH_3CHOHCN$	$C_6H_5CHOHCN$
$NaHSO_3$(aq), conc forms white solids	$CH_3CHOHSO_3Na$	$C_6H_5CHOHSO_3Na$
2,4-Dinitrophenyl-hydrazine forms solid hydrazones	$CH_3CH{=}NNHC_6H_3(NO_2)_2$	$C_6H_5CH{=}NNHC_6H_5(NO_2)_2$
Hydroxylamine, NH_2OH, forms solid oximes	$CH_3CH{=}NOH$	$C_6H_5CH{=}NOH$
PCl_5	CH_3CHCl_2	$C_6H_5CHCl_2$

Differences between aldehydes and ketones

Reagent/reaction	*Ethanal, CH_3CHO*	*Propanone, CH_3COCH_3*
Polymerization. NaOH(aq) or H_2SO_4	Very readily to dimer, trimer and tetramer	Dimer only under special conditions (other ketones do not polymerize)
Fehling's solution Cu^{2+} complex ion	Red precipitate, Cu_2O	No reaction
Tollen's reagent, $[Ag(NH_3)_2]^+$	Silver mirror	No reaction
Oxidation, MnO_4^-/H^+	Readily to CH_3COOH. Same number of C atoms	More difficult to CH_3COOH. Fewer C atoms
Reduction. H_2/Ni or $Li[AlH_4]$ (ether) or $Na[BH_4]$ (methanol-water)	Primary alcohol, CH_3CH_2OH ethanol	Secondary alcohol, $CH_3CHOHCH_3$ propan-2-ol
NH_3(ether)	Addition. $CH_3CHOHNH_2$ ethanal-ammonia	Very complex compound

Special reactions of methanal, HCHO

It is the strongest reducing agent of the aldehydes, and reduces $HgCl_2$(aq) to white Hg_2Cl_2(s) and then to Hg. Readily polymerizes, forming a white solid poly(methanal) when evaporated. NH_3(aq) forms a complex compound (compare benzaldehyde); other aldehydes form addition products.

Uses of aldehydes and ketones

Methanal preserves biological specimens. With phenol it forms the plastic Bakelite, and with urea another plastic. Benzaldehyde is in some flavours added to foods. Propanone is a solvent for ethyne in storage cylinders, and is used to make trichloromethane (a solvent), and triiodomethane (an antiseptic).

QUESTIONS

1. Mention two reactions in which ethanal reacts as an unsaturated compound and two reactions in which it reduces inorganic compounds.

2. The ***condensation*** reactions of propanone with certain compounds may be regarded as ***addition*** reactions followed by ***elimination*** reactions. Name three of the 'certain compounds' and write equations for the reactions. Explain clearly what is meant by the three words in italics.

3. Mention three reactions in which ethanal and propanone resemble each other and three reactions in which they differ. Account for the differences.

4. Compare and contrast methanal, ethanal and benzaldehyde by reference to (a) their physical states, (b) their reactions with (i) oxidizing agents, (ii) ammonia, and (iii) sodium hydroxide. Account briefly for the differences.

5. A liquid (b.p. 453 K) is insoluble in water and is considered to be an aldehyde. Describe chemical tests by which you would determine whether or not it is an aldehyde.

6. You are required to prepare ethanal in large quantities in the laboratory. You have available the following starting materials: ethanol, ethanoyl chloride, a mixture of equimolar proportions of calcium methanoate (formate) and calcium ethanoate (acetate), and 1,1-dichloroethane. Outline, with essential practical details, how, from each of the four substances, reasonably pure ethanal could be prepared. (No diagrams are required.) Which method would you recommend as being the one which would give the highest percentage yield of pure, dry ethanal? Give reasons why other methods would give poorer percentage yields. (L.)

7. Comment on or explain the following: (a) the H—C—H bond in methane is 109.5°, whilst the H—C—H bond in ethene is 120°; (b) ethanol boils at 351 K, but its isomer, methoxyethane, boils at 248 K; (c) ethanol boils at 351 K but the compound of formula C_2H_5SH boils at 309 K; (d) methanal is a stronger reducing agent than ethanal. (C.)

8. Ozonolysis of compound ***A***(C_6H_{12}) gave neutral compounds ***B***(C_3H_6O) and ***C***(C_3H_6O). ***B*** did not reduce Fehling's solution but ***C*** did. Reduction of ***C*** with hydrogen and a catalyst gave ***D*** (C_3H_8O), which, when heated with concentrated hydrobromic acid, gave ***E*** (C_3H_7Br). ***E*** was heated with a concentrated solution of potassium hydroxide in ethanol, and gave ***F*** (C_3H_6). Identify the compounds ***A-F***, explaining your reasoning. (C.)

9. Name the isomers which have the formula C_7H_7Cl and write their structural formulae. One of the isomers ***G***, when refluxed with potassium carbonate solution, forms ***H***, formula C_7H_8O. Identify ***G*** and describe how it can be

converted into pure *J*, formula C_7H_6O. How does *J* react with (a) ammonia, (b) potassium hydroxide? How can *J* be converted to *K*, formula $C_{13}H_{12}N_2$?

(C.)

10. Butan-2-one and pent-1-ene both contain a double bond. (a) How do you account for the difference of 50 K between the boiling points of these compounds although their relative molecular masses are about the same? (b) State one reagent that will form an addition product with the ketone but not with the alkene and one reagent that will form an addition compound with the alkene but not with the ketone. Write equations for the reactions involved. (c) Indicate possible mechanisms for the reactions in (b) and show clearly the differences in the two.

11. Three carbonyl-containing compounds have the following composition by mass: C = 66.7 per cent; H = 11.1 per cent; O = 22.2 per cent. If their relative molecular mass is 72, find their empirical and molecular formulae, and write the three structural formulae. An organic compound *A*, $C_2H_4O_3$, can be oxidized first to *B*, $C_2H_2O_3$, and then to *C*, $C_2H_2O_4$. Suggest possible structural formulae of *A*, *B* and *C*.

Chapter 12
CARBOHYDRATES

Carbohydrates have the general formula $C_nH_{2m}O_m$. The name 'carbohydrates' is not a suitable one because they cannot be formed from carbon and water and they are not hydrates $C_n(H_2O)_m$.

The compounds are classified into three groups according to whether or not they can be hydrolysed and to the products of hydrolysis.

Monosaccharides include glucose and fructose, which are isomers of formula $C_6H_{12}O_6$. They do not hydrolyse. Glucose is an aldehyde and is an *aldose*; fructose is a ketone and is a *ketose.* Since their molecules contain six carbon atoms, glucose and fructose are *hexoses.* By combining the various names, glucose is an *aldohexose* and fructose is a *ketohexose.*

Disaccharides are carbohydrates which on hydrolysis give two monosaccharide units. Examples are sucrose, maltose and lactose, which are isomers of formula $C_{12}H_{22}O_{11}$.

$$C_{12}H_{22}O_{11} + H_2O \xrightarrow{H^+(aq)} C_6H_{12}O_6 + C_6H_{12}O_6$$

Hydrolysis of sucrose produces glucose and fructose, and hydrolysis of maltose produces glucose only.

Polysaccharides are carbohydrates which on hydrolysis produce many monosaccharide units from one molecule. They are polymers and their relative molecular masses are high. Starch molecules consist of as many as 3000 glucose units (combined together with loss of water) and cellulose molecules can contain 5000 glucose units and have a relative molecular mass of about 1 000 000.

$$(C_6H_{10}O_5)_n + nH_2O \xrightarrow{H^+(aq)} nC_6H_{12}O_6$$

Glycogen, the food reserve stored in the livers of many animals, is a polysaccharide.

Occurrence of carbohydrates

Glucose (grape sugar, blood sugar) and fructose (fruit sugar) are in the juices of many fruits, and in nectar and honey. Glucose is in the leaves and roots of many plants and is always in the blood of animals.

Sucrose (cane sugar, beet sugar) is in the stems of sugar-cane and in sugar-beet. The stems of sugar-cane are squeezed between metal rollers to extract the juice; the roots of sugar-beet are shredded and sucrose is extracted by hot water. After treatment to remove impurities, a yellow or brown 'raw sugar' is obtained. Charcoal is used to absorb the colouring matter and to produce white sugar.

Maltose is the intermediate product formed when starch is hydrolysed by the enzyme diastase during the fermentation of starch to ethanol (p. 119). Lactose (milk sugar) is present in milk.

Starch is the final product of photosynthesis in the green leaves of plants. Plants store starch so that they can survive even when photosynthesis is no longer possible. The grains of wheat, barley, oats, maize and rice contain starch; stem tubers of potatoes and yams and root tubers of sweet-potatoes and cassava contain much starch.

Cellulose is the chief constituent of the walls of plant cells. Cotton and flax (for linen) contain about 80 per cent by mass of cellulose, hemp and jute (for ropes, string, sailcloth, and so on) contain about 70 per cent, and wood pulp, straw, and esparto grass (for making paper) contain 20 to 40 per cent. Filter paper of good quality is almost pure cellulose.

Fehling's solution

These solutions are used in tests for sugars, aldehydes and ketones.

Fehling's solution. The solution is not stable and therefore it is usual to make two separate solutions and mix them only when required for use. (a) Dissolve about 70 g of copper sulphate-5-water in 1 litre of water. (b) Dissolve 120 g of sodium hydroxide and 350 g of sodium potassium 2,3-dihydroxybutanedioate (sodium potassium tartrate) in 1 litre of water. A mixture of equal volumes of the solutions is Fehling's solution.

Reactions of carbohydrates

[Prepare starch solution by mixing about 1 g of starch with 10 cm^3 of water and then pouring the paste into about 100 cm^3 of boiling water. Boil for two minutes and stir vigorously. A crystal of mercury(II) iodide or 2-hydroxybenzoic acid (salicylic acid) prevents the growth of moulds.]

1. *Concentrated sulphuric acid.*
(a) Add a few cm^3 of the acid to about 1 g of glucose in a dry test-tube. Note any reaction. Warm gently, and use an asbestos square to protect the bench top in case mixture comes out of the tube.

(b) Repeat the test with fructose, sucrose, starch and cellulose.

(c) Cover the bottom of a 500 cm^3 beaker with glucose and moisten the sugar with water. Stand the beaker on asbestos. Carefully add concentrated sulphuric acid to a depth of about 1 cm. Observe all the changes.

2. *Iodine test for starch.* Add one drop of a solution of iodine in aqueous potassium iodide to dilute starch solution. Observe the colour. Warm the mixture and then allow to cool. What happens?

3. *Fehling's solution* (Cu^{2+} as oxidizing agent). Add a few glucose crystals to about 10 cm^3 of the solution and warm. Observe the changes.

Repeat the test with other carbohydrates.

4. *Tollen's reagent* ($[Ag(NH_3)_2]^+$ as oxidizing agent, p. 149). Add glucose solution (about 2 per cent) to Tollen's reagent in a clean test-tube. Place the tube in a beaker of warm water. Observe any changes.

Repeat with the other carbohydrates.

5. *Phenylhydrazine.* Mix phenylhydrazine (0.5 cm^3), glacial ethanoic acid (0.5 cm^3), and sodium ethanoate crystals (about 1 g). Dissolve about 0.5 g of glucose in 10 cm^3 of water and add the glucose solution to the mixture. Heat the tube and contents in a beaker of boiling water for about 20 minutes. Observe the crystals that form.

Repeat with the other carbohydrates, including lactose and maltose if available. Allow the hot solution to cool if no precipitate forms.

6. *Hydrolysis of sucrose.* Boil aqueous sucrose with dilute hydrochloric acid for 5 minutes. Allow to cool and just neutralize with aqueous sodium hydroxide. Test the product with (a) Fehling's solution, (b) Tollen's reagent, and (c) phenylhydrazine (none of these reagents reacts with sucrose).

7. *Hydrolysis of starch.*
(a) Boil aqueous starch solution with dilute hydrochloric acid for 5 minutes, cool, neutralize with aqueous sodium hydroxide, and test the product with Fehling's solution (which is unaffected by starch) and with iodine solution.

(b) Add about 10 cm^3 of dilute hydrochloric acid to about 30 cm^3 of starch solution. Divide the mixture into four test-tubes. Add one drop of iodine solution to the first tube. Place the other three tubes in a beaker of boiling water. Remove them after about 5, 10 and 15 minutes respectively, cool the contents by adding cold water, and then add one drop of iodine solution.

(c) *Hydrolysis by enzymes.* To a few cm^3 of starch solution add a saltspoonful of diastase (or use saliva from the mouth if diastase is not available; saliva contains the enzyme ptyalin). Warm the mixture to about 310 K, which is body temperature. After one minute use a glass rod to remove one drop of the mixture and add it to one drop of iodine solution on a white tile. Repeat at intervals of one minute. Finally, test the product with Fehling's solution.

Repeat the test with diastase or saliva in alkaline solution (use a few cm^3 of 1 per cent sodium carbonate solution) and in acid solution (use a few cm^3 of dilute hydrochloric acid). How do the acid and alkali affect the rate of hydrolysis of the starch by the enzyme?

Physical properties of carbohydrates

Glucose-1-water, $C_6H_{12}O_6 \cdot H_2O$, is a white crystalline solid with a sweet taste. It is very soluble in water but sparingly soluble in ethanol. The hydrate melts at 359 K and the anhydrous sugar at 419 K.

Fructose is a white crystalline anhydrous solid, m.p. 368 K. It is sweeter than glucose, more soluble in water, and slightly soluble in ethanol.

Sucrose is a white crystalline solid, m.p. 433 K. It is readily soluble in water but only slightly soluble in ethanol. After melting it solidifies on cooling to a brown transparent solid called 'barley sugar'. It loses some water when heated at about 470 K and forms brownish caramel, used as a colouring substance for gravies, rum, beer, and so on.

Starch is a non-crystalline tasteless solid. It is insoluble in cold water. Boiling water bursts the granules, and the starch then forms a colloidal solution or *sol* which sets to a *gel* (starch paste) on cooling, provided the sol is not too dilute. 'Soluble starch' is obtained as a precipitate by adding ethanol to a starch sol; the solid readily forms a colloidal solution with water.

Cellulose is an amorphous solid. It is insoluble in water, ethanol and ether, but dissolves in ammoniacal copper(II) sulphate.

Glucose, $C_6H_{12}O_6$

1. *Reduction.* Hydrogen under pressure in the presence of a nickel catalyst converts glucose to hexane. Phosphorus and hydrogen iodide convert glucose to 1-iodohexane and 2-iodohexane. The three reduction products have unbranched carbon chains.

$$C_6H_{12}O_6 \xrightarrow{H_2/Ni} \underset{\text{hexane}}{CH_3CH_2CH_2CH_2CH_2CH_3}$$

$$C_6H_{12}O_6 \xrightarrow{HI/P} \underset{\text{1-iodohexane}}{CH_3CH_2CH_2CH_2CH_2CH_2I}$$

2. *Ethanoic anhydride.* This reagent converts every hydroxyl group to an ethanoate group. The product is a pentaethanoate, indicating five OH groups in a glucose molecule.

$$C_6H_7O(OH)_5 \xrightarrow{(CH_3CO)_2O} C_6H_7O(OOCCH_3)_5$$

3. *Iodomethane.* Glucose is methylated by treatment with iodomethane and moist silver oxide. The product, a pentamethyl derivative, confirms the presence of five hydroxyl groups:

$$C_6H_7O(OH)_5 \xrightarrow{CH_3I} C_6H_7O(OCH_3)_5$$

4. *Aldehyde properties of glucose.*

(a) *Reduction.* Sodium amalgam and water convert glucose to a primary alcohol:

$$C_5H_{11}O_5CHO \xrightarrow{Na/Hg} C_5H_{11}O_5CH_2OH$$

(b) *Oxidation.* Fehling's solution and Tollen's reagent oxidize glucose to an acid. Reddish-brown copper(I) oxide is precipitated from Fehling's solution:

$$C_5H_{11}O_5CHO + 2Cu^{2+} + 5OH^- \rightarrow Cu_2O(s) + 3H_2O + C_5H_{11}O_5COO^-$$

A silver mirror is deposited on the test-tube when glucose reduces Tollen's reagent (diamminesilver, $[Ag(NH_3)_2]^+$):

$$C_5H_{11}O_5CHO + 2Ag^+ + 3OH^- \rightarrow 2Ag(s) + 2H_2O + C_5H_{11}O_5COO^-$$

(c) *Addition product.* Glucose forms an addition product with hydrogen cyanide, but not with either ammonia or sodium hydrogensulphite.

$$C_5H_{11}O_5CHO + HCN \rightarrow \underset{\text{a hydroxynitrile}}{C_5H_{11}O_5CH(OH)CN}$$

(d) *Condensation or addition-elimination reactions.* Glucose forms an oxime with hydroxylamine:

$$C_5H_{11}O_5CHO + H_2NOH \rightarrow H_2O + \underset{\text{an oxime}}{C_5H_{11}O_5CH{=}NOH}$$

Phenylhydrazine first forms a phenylhydrazone:

$$C_5H_{11}O_5CHO + H_2NNHC_6H_5 \rightarrow H_2O + \underset{\text{a phenylhydrazone}}{C_5H_{11}O_5CH{=}NNHC_6H_5}$$

Further reaction occurs because the CHOH group next to the aldehyde group is oxidized by phenylhydrazine to a carbonyl group (p. 172), which also reacts to form a phenylhydrazone. The product has two adjacent phenylhydrazone residues and is called an *osazone*:

$$\begin{array}{c} CH_2OH \\ | \\ (CHOH)_3 \\ | \\ CHOH \\ | \\ CHO \end{array} \xrightarrow{C_6H_5NHNH_2} \begin{array}{l} CH_2OH \\ | \\ (CHOH)_3 \\ | \\ C{=}NNHC_6H_5 \\ | \\ CH{=}NNHC_6H_5 \end{array}$$

5. *Dehydration.* Concentrated sulphuric acid dehydrates glucose (and all other sugars) to form carbon (sugar charcoal):

$$C_6H_{12}O_6 \rightarrow 6C(s) + 6H_2O$$

6. *Fermentation.* Ethanol is produced (p. 119).

In the above reactions glucose seems to be an unbranched-chain pentahydroxy-aldehyde 2, 3, 4, 5, 6-pentahydroxyhexanal:

$$CH_2OHCHOHCHOHCHOHCHOHCHO \quad \text{or} \quad CH_2OH(CHOH)_4CHO$$

Since glucose does not react with ammonia or sodium hydrogensulphite, its aldehyde group is not the same as that in true aldehydes. Moreover, with dry hydrogen chloride as a catalyst, an aldehyde reacts with two molecules of ethanol to form a diethoxy-compound:

$$\begin{array}{c} H \\ | \\ R{-}C{=}O \end{array} \xrightarrow{C_2H_5OH} \underset{\text{a hemiacetal}}{\begin{array}{c} H \\ | \\ R{-}C{-}OH \\ | \\ OC_2H_5 \end{array}} \xrightarrow[(-H_2O)]{C_2H_5OH} \underset{\text{an acetal}}{\begin{array}{c} H \\ | \\ R{-}C{-}OC_2H_5 \\ | \\ OC_2H_5 \end{array}}$$

However, glucose reacts with only one molecule of ethanol to form an acetal, and therefore glucose must be a hemiacetal. One of the hydroxyl groups reacts with the aldehyde group to form a cyclic compound. A glucose molecule has a ring of five carbon atoms and one oxygen atom (p. 173). However, the ring is easily broken by hydroxylamine and phenylhydrazine though not by weaker reagents such as ammonia and sodium hydrogensulphite.

Fructose, $C_6H_{12}O_6$

Fructose is an isomer of glucose and some of its properties are similar. For example, with ethanoic anhydride fructose forms a pentaethanoate, indicating that there are five OH groups in a fructose molecule. A carbonyl group is indicated by the addition product formed with hydrogen cyanide, and by the reduction of Fehling's solution and Tollen's reagent. Fructose forms an oxime with hydroxylamine, and with phenylhydrazine it forms the same osazone as that formed with glucose.

In the above reactions fructose seems to be a pentahydroxyketone:

$$CH_2OHCHOHCHOHCHOHCOCH_2OH \quad \text{or} \quad CH_2OH(CHOH)_3COCH_2OH$$

Therefore glucose and fructose differ only in the two end groups:

$$\begin{array}{c} CH_2OH \\ | \\ (CHOH)_3 \\ | \\ CHOH \\ | \\ CHO \\ \text{glucose} \end{array} \qquad\qquad \begin{array}{c} CH_2OH \\ | \\ (CHOH)_3 \\ | \\ C{=}O \\ | \\ CH_2OH \\ \text{fructose} \end{array}$$

The reducing reactions of fructose are due to the groups CHOHCO. Sodium bromate(I), NaBrO, oxidizes glucose to an acid, $C_6H_{12}O_7$, but has no action on fructose, indicating the absence of an aldehyde group.

Equations for the reaction with phenylhydrazine are:

$$\begin{array}{c} CH_2OH \\ | \\ (CHOH)_3 \\ | \\ C{=}O \\ | \\ CH_2OH \end{array} \xrightarrow{C_6H_5NHNH_2} \begin{array}{c} CH_2OH \\ | \\ (CHOH)_3 \\ | \\ C{=}NNHC_6H_5 \\ | \\ CH_2OH \\ \text{a phenylhydrazone} \end{array} \rightarrow \begin{array}{c} CH_2OH \\ | \\ (CHOH)_3 \\ | \\ C{=}NNHC_6H_5 \\ | \\ CH{=}NNHC_6H_5 \\ \text{an osazone} \end{array}$$

Fructose phenylhydrazone differs from glucose phenylhydrazone but the osazone is the same. In its reactions with both sugars, phenylhydrazine oxidizes the CHOH group next to the phenylhydrazone group:

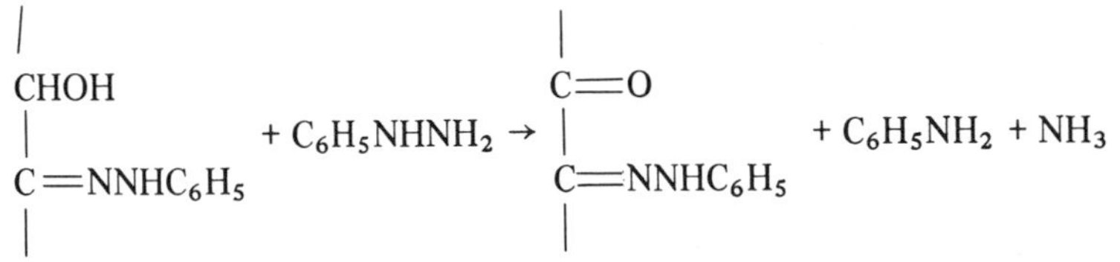

$$\begin{array}{l} | \\ \mathrm{CHOH} \\ | \\ \mathrm{C{=}NNHC_6H_5} \\ | \end{array} + \mathrm{C_6H_5NHNH_2} \rightarrow \begin{array}{l} | \\ \mathrm{C{=}O} \\ | \\ \mathrm{C{=}NNHC_6H_5} \\ | \end{array} + \mathrm{C_6H_5NH_2} + \mathrm{NH_3}$$

A fructose molecule has a cyclic structure as below; compare it with that of glucose. (Do not learn these configurations.)

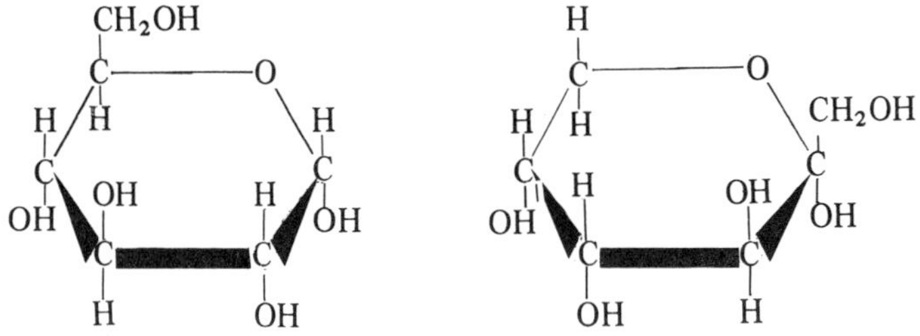

glucose

fructose

Uses of starch and cellulose

Starch is used in the manufacture of ethanol and glucose, for stiffening linen, sizing paper, and as an adhesive.

Cellulose is used in the manufacture of paper (from wood, straw, cotton waste, rags, esparto grass). It is the chief compound in cotton, linen, hemp and jute. Cellulose nitrate (gun cotton) is a powerful explosive because a little solid forms a large volume of gaseous products, e.g.

$$2[C_6H_7O_2(NO_3)_3]_n \rightarrow 3nCO_2 + 9nCO + 7nH_2O + 3nN_2$$

Lacquer paints (used on cars), celluloid and some artificial leather are made from partially nitrated celluloses.

Cellulose ethanoates (acetates) are used to make rayon (artificial silk) and lacquers.

SUMMARY

Carbohydrates

General formula: $C_nH_{2m}O_m$. Monosaccharides: glucose and fructose, $C_6H_{12}O_6$. Disaccharides: sucrose, maltose, lactose, $C_{12}H_{22}O_{11}$. Polysaccharides: starch, cellulose, glycogen, $(C_6H_{10}O_5)_n$.

Glucose and fructose, $C_6H_{12}O_6$

White crystalline solids. Sweet taste. Soluble in water.

Reagent/reaction	*Glucose (an aldose)* $CH_2OH(CHOH)_4CHO$	*Fructose (a ketose)* $CH_2OH(CHOH)_3COCH_2OH$
Reduction. H_2/Ni	Hexane, $CH_3(CH_2)_4CH_3$	Hexane
Ethanoylation. $(CH_3CO)_2O$	Forms a pentaethanoate	Forms a pentaethanoate
Reduction. Na/Hg	Primary alcohol	Secondary alcohol
Oxidation		
Fehling's solution	Red precipitate, Cu_2O	Red precipitate, Cu_2O
Tollen's reagent	Silver mirror	Silver mirror
$Br_2(aq)$	Decolorizes	No action
Addition		
HCN	Hydroxynitrile	Hydroxynitrile
Hydroxylamine	Oxime	Oxime
H_2SO_4 (conc)	Dehydrates to carbon	Dehydrates to carbon

Glucose differs from aldehydes because it does not form addition products with NH_3 or $NaHSO_3$. Fructose differs from ketones because it reacts with Fehling's solution and Tollen's reagent.

QUESTIONS

1. Mention two reactions in which glucose behaves as an aldehyde and two chemical ways in which it does not resemble ethanal. Explain the differences by reference to the structure of glucose.

2. Describe reactions of fructose which indicate that it is a ketone. Explain why a fructose molecule does not contain a carbonyl group.

3. Describe the reactions of (a) glucose, and (b) fructose with (i) hydroxylamine, and (ii) phenylhydrazine.

Chapter 13
CARBOXYLIC ACIDS

Carboxylic acids contain one or more *carboxyl groups*,

$$—COOH \quad \text{or} \quad —C\begin{matrix} {}^{\nearrow\!\!\!\nearrow}O \\ {}_{\searrow}OH \end{matrix}$$

A carboxyl group seems to contain a carbonyl group (present in aldehydes and ketones) and a hydroxyl group (present in alcohols). However, the two groups influence and change the properties of each other.

Name	*Formula*	*M.p./K*	*B.p./K*	*Density/g cm^{-3}*
Methanoic (formic) acid	$HCOOH$	281.5	374	1.22
Ethanoic (acetic) acid	CH_3COOH	290	391	1.05
Propanoic acid	CH_3CH_2COOH	252	414	0.99
Butanoic acid	$CH_3CH_2CH_2COOH$	269	437	0.96
2-Methylpropanoic acid	$(CH_3)_2CHCOOH$	226	427.5	0.95
Benzoic acid	C_6H_5COOH	395	522	1.32
Ethanedioic acid	$HOOCCOOH$	430	462.6	1.65

Monocarboxylic aliphatic acids may be regarded as derivatives of alkanes. The general formula is $C_nH_{2n}O_2$ or C_nH_{2n+1} COOH, and $n = 0, 1, 2$, and so on.

They are called *alkanoic* acids, and were formerly called *fatty acids* because animal and vegetable fats are esters of many of them. The first two acids are methanoic acid (formic acid), $HCOOH$ and ethanoic acid (acetic acid), CH_3COOH. Benzoic acid (benzenecarboxylic acid), C_6H_5COOH, is the simplest aromatic acid.

Alkanoic acids are isomeric with esters, e.g.

Ethanoic acid, CH_3COOH, and methyl methanoate, $HCOOCH_3$;
Propanoic acid, C_2H_5COOH, methyl ethanoate, CH_3COOCH_3, and ethyl methanoate, $HCOOC_2H_5$.

Occurrence of acids

Methanoic acid was first obtained from red ants (in 1749). Ethanoic acid is in vinegar, which is a 4 per cent aqueous solution formed when dilute ethanol (beer and some wines) turns sour in air.

The cells of animals and vegetables contain fats and oils, which are esters of the alcohol propane-1,2,3-triol, $CH_2OHCHOHCH_2OH$ (glycerol), with certain carboxylic acids, all of which contain an even number of carbon atoms, e.g.

Acid	*Formula*	*Fat or oil*
Octanoic	$CH_3(CH_2)_6COOH$	Coconut oil
Decanoic	$CH_3(CH_2)_8COOH$	Coconut oil
Hexadecanoic (palmitic)	$CH_3(CH_2)_{14}COOH$	Palm oil
Octadecanoic (stearic)	$CH_3(CH_2)_{16}COOH$	Animal fats
Octadec-9-enoic (oleic)	$C_{17}H_{33}COOH$	Olive oil

The last acid is unsaturated: $CH_3(CH_2)_7CH{=}CH(CH_2)_7COOH$.

Butter contains an ester of butanoic acid. Benzoic acids occur in many resins and in the urine of horses and other herbivorous animals. Ethanedioic acid occurs in the roots and leaves of rhubarb as its salt $COOHCOOK$.

Preparation of carboxylic acids

1. *Oxidation of primary alcohols or aldehydes.* The oxidizing agent is acidified sodium dichromate(VI), potassium dichromate(VI) or potassium manganate(VII). Dilute sulphuric acid is the acid used.

$$RCH_2OH \rightarrow RCHO \rightarrow RCOOH$$

$$\underset{}{CH_3CH_2OH} \rightarrow \underset{\text{ethanal}}{CH_3CHO} \rightarrow \underset{\text{ethanoic acid}}{CH_3COOH}$$

On a commercial scale, air is the oxidizing agent and various catalysts are used. Methanal vapour and air form methanoic acid when passed over heated platinum as a catalyst (a laboratory method). Alkaline potassium manganate(VII) is used to oxidize phenylmethanol:

$$C_6H_5CH_2OH \xrightarrow{MnO_4^-/OH^-} C_6H_5CHO \rightarrow \underset{\text{benzoic acid}}{C_6H_5COOH}$$

2. *Hydrolysis of nitriles.* Heat the nitrile under reflux with a mineral acid or aqueous sodium (or potassium) hydroxide. The catalytic action of the acid is greater than that of alkali:

$$RCN \xrightarrow{H_3O^+ \text{ or } OH^-} RCONH_2 \rightarrow RCOONH_4 \rightarrow RCOOH \text{ or } RCOONa$$

$$\underset{\text{ethanenitrile}}{CH_3CN} + 2H_2O \rightarrow NH_3 + CH_3COOH$$

$$\underset{\text{benzonitrile}}{C_6H_5CN} + 2H_2O \rightarrow NH_3 + C_6H_5COOH$$

An amide (p. 244) is the first product, followed by the ammonium salt of the acid. Acidic hydrolysis decomposes the ammonium salt and forms the free acid; alkaline hydrolysis liberates ammonia gas and leaves the sodium (or potassium) salt.

3. *Oxidation of aromatic hydrocarbons.* Use dilute nitric acid or alkaline potassium manganate(VII) to oxidize arenes with a side-chain, which forms a carboxyl group:

$$\underset{\text{methylbenzene}}{C_6H_5CH_3(l)} \rightarrow C_6H_5COOH(s)$$

$$\underset{\text{ethylbenzene}}{C_6H_5C_2H_5(l)} \rightarrow C_6H_5COOH(s)$$

To prepare ethanoic (acetic) acid from ethanol

Principle. Ethanol is oxidized by heating under reflux with excess oxidizing agent.

Method. Arrange a reflux-distillation apparatus as on p. 78.

Add concentrated sulphuric acid (9 cm^3) carefully to cold water (10 cm^3) in the small (50 cm^3) flask of Fig. 5.2(b). Then add sodium dichromate(VI) (11 g) to the mixture. Attach the cold-finger condenser and pass a fast stream of cold water.

Add water (8 cm^3) to ethanol (4 cm^3) in a test-tube. Use a teat pipette to add this aqueous ethanol, 1 cm^3 at a time, down the inside of the condenser. Allow about one minute for each portion of ethanol to be oxidized before adding the next portion; the addition takes at least 10 minutes. Finally, heat the mixture gently under reflux for about 20 minutes to ensure complete oxidation.

Rearrange the condenser for ordinary distillation. Distil the mixture and collect about 15 cm^3 of distillate in a test-tube. This is aqueous ethanoic acid, in which the main impurities are ethanol and possibly ethanal.

(Larger quantities may be used if larger apparatus is used.)

To prepare benzoic acid from methylbenzene

Principle. The methyl side-chain in the arene is oxidized to the carboxyl group by alkaline potassium manganate(VII):

$$C_6H_5CH_3(l) + 3[O] \rightarrow C_6H_5COOH(s) + H_2O$$

$$C_6H_5CH_3(l) + 2MnO_4^-(aq) \rightarrow C_6H_5COOH(s) + 2MnO_2(s) + 2OH^-(aq)$$

Method. Arrange a reflux distillation apparatus as on p. 78.

To the flask add methylbenzene (3 cm^3), potassium manganate(VII) (10 g) and dilute sodium hydroxide (20 cm^3). Attach the cold-finger condenser. Heat the mixture under reflux until the oily methylbenzene has disappeared; this takes 3 to 4 hours.

Allow the mixture to cool, filter off the precipitated manganese(IV) oxide, and acidify the filtrate with concentrated hydrochloric acid. Benzoic acid is precipitated as a white crystalline solid. Either (a) filter and recrystallize the acid from hot water, or (b) extract the acid with ether and then remove the ether by distillation on a water bath. The acid may be purified by sublimation.

Reactions of alkanoic acids and benzoic acid

1. *Solubility.* Add water, drop by drop, to ethanoic acid in a test-tube.

Add a saltspoonful of benzoic acid crystals to a test-tube half filled with cold water. Shake well. Boil to obtain a solution, adding more water if necessary. Allow to cool.

2. *Acidity.* Add litmus and universal indicator to separate portions of aqueous ethanoic acid and then to hot aqueous benzoic acid.

3. *Sodium hydroxide.* Use a teat pipette to add ethanoic acid to a few cm^3 of aqueous sodium hydroxide in a basin. Add the acid until litmus paper just turns red. Evaporate the solution to obtain crystals of sodium ethanoate-3-water, $CH_3COONa \cdot 3H_2O$.

4. *Sodium carbonate and hydrogencarbonate.* Add solid sodium carbonate and solid sodium hydrogencarbonate to separate portions of aqueous ethanoic acid.

Repeat the test with a hot solution of benzoic acid.

5. *Metals.* Add clean magnesium ribbon to aqueous ethanoic acid. Identify any gas formed.

Add tiny pieces of clean sodium to about 1 cm^3 of glacial ethanoic acid in a dry basin.

6. *Ester formation.*
(a) Repeat test 4 on p. 112 (using ethanoic acid, ethanol and concentrated sulphuric acid).

(b) Repeat with pentan-1-ol instead of ethanol. Observe the smell of ethyl pentanoate.

(c) Repeat with benzoic acid instead of ethanoic acid.

7. *Heat.* Heat benzoic acid in an ignition tube.

8. *Sodalime.* Grind together equal volumes of benzoic acid and sodalime. Heat the mixture in an ignition tube. Smell the product, ignite the vapour, and observe the flame.

Sodium ethanoate and sodalime form methane (p. 18).

9. *Phosphorus pentachloride.* Add a little pentachloride to about 1 cm^3 of glacial ethanoic acid in a dry basin.

Repeat the test with benzoic acid crystals.

10. *Iron(III) chloride.* Either carefully neutralize aqueous ethanoic acid with sodium hydroxide or make a solution of sodium ethanoate (acetate). Make iron(III) chloride solution almost neutral by adding dilute sodium hydroxide drop by drop until a faint permanent precipitate of brownish iron(III) hydroxide is formed; filter off the precipitate.

Mix the two solutions, and observe the colour. Divide the product into two portions. Add dilute hydrochloric acid to one portion, and boil the second portion. Observe any changes.

Repeat the test with sodium methanoate and with ammonium benzoate.

Special reactions of methanoic (formic) acid and ethanedioic acid

1. *Tollen's reagent* (*diamminesilver*, $[Ag(NH_3)_2]^+$). Use sodium methanoate instead of the aldehyde in test 5 on p. 149.

2. *Fehling's solution.* Use methanoic acid instead of the aldehyde in test 6 on p. 149.

3. *Potassium manganate*(VII) *and potassium dichromate*(VI). Acidify methanoic acid or sodium methanoate with dilute sulphuric acid. Add cold dilute potassium manganate(VII), drop by drop. Observe any change.

Repeat the test with potassium dichromate(VI) in place of the manganate(VII) solution.

Repeat the tests with ethanedioic acid, $(COOH)_2$, instead of methanoic acid.

4. *Mercury*(II) *chloride.* Add a few drops of the aqueous chloride solution to a few cm^3 of sodium methanoate. Note the first and final colours of the precipitates.

5. *Concentrated sulphuric acid.* Add a few cm^3 of concentrated sulphuric acid to sodium methanoate or methanoic acid in a test-tube. Warm gently if necessary. Ignite the gas evolved and observe the flame. Note if charring occurs.

Repeat the test with a few crystals of ethanedioic acid.

Manufacture of carboxylic acids

Air under pressure and in the presence of a catalyst is used to oxidize ethanal to ethanoic acid. Ethanal is manufactured from ethene (p. 48).

A mixture of pentane, hexane and heptane (obtained from petroleum) is oxidized by air at a high temperature and pressure. Methanoic, ethanoic and propanoic acids are present in the product and they are separated by distillation.

Beers and wines with less than 15 per cent ethanol become sour in air because bacterial oxidation produces ethanoic acid. The product with about 4 per cent acid is called vinegar. In the *Quick Vinegar Process* dilute aqueous ethanol passes slowly down a tall tower packed with wood shavings that have been soaked in vinegar (to provide the necessary bacteria). The shavings have a large surface area, so that there is plenty of contact between air and ethanol, and they are a good medium for growth of the bacteria. Warm air is blown up the tower.

Hexane-1,6-dioic acid. $HOOC(CH_4)_2COOH$, is used in the manufacture of nylon 66. Cyclohexanol is made by (a) reducing phenol with hydrogen using a nickel catalyst, or (b) reducing benzene to cyclohexane and oxidizing this with air. Oxidation of cyclohexanol yields the acid:

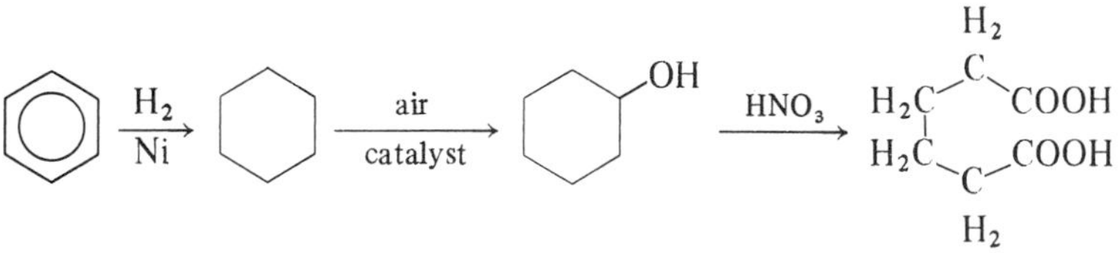

$$\text{benzene} \xrightarrow[\text{Ni}]{H_2} \text{cyclohexane} \xrightarrow[\text{catalyst}]{\text{air}} \text{cyclohexanol} \xrightarrow{HNO_3} \text{hexane-1,6-dioic acid}$$

Properties of carboxylic acids

Physical properties

Methanoic acid is a colourless liquid with a pungent smell. It irritates and destroys skin. It is denser than water and forms crystals when frozen. It is completely miscible with water, forming an acid solution; the degree of ionization in a 1 M solution is 1.5 per cent. In the solid state and in non-ionizing solvents (e.g. benzene), the acid exists as the dimer $(HCOOH)_2$, (see next paragraph).

Ethanoic acid is a colourless liquid with a sharp smell of vinegar and a sour taste. In cold weather, it forms an ice-like solid and therefore the pure acid is called *glacial* ethanoic acid. The acid is hygroscopic and completely miscible with water; the degree of ionization in a 1 M solution is 0.4 per cent. The pure acid and a concentrated aqueous solution burn and blister skin; they are used to remove warts from the body. In the vapour state and in benzene solution, the acid exists as the dimer $(CH_3COOH)_2$. Hydrogen bonds (represented by four dots) exist between two molecules:

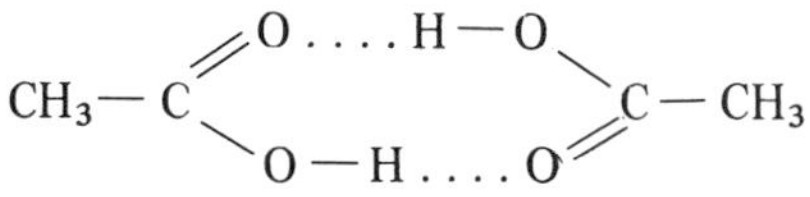

$$CH_3-C\begin{matrix}\nearrow\!\!\!\!\!= O \;....\; H-O \searrow \\ \searrow O-H \;....\; O =\!\!\!\!\!\nearrow\end{matrix}C-CH_3$$

Propanoic and *butanoic acids* are completely miscible with water but *2-methylpropanoic acid*, $(CH_3)_2CHCOOH$, is not, and 100 g of water at 293 K dissolve only 20 g of the acid.

Benzoic acid is a white crystalline solid. It sublimes readily at about 373 K to form snow-like needles. It is only sparingly soluble in cold water (about 0.3 g per 100 g of water) but it forms a 6 per cent solution with boiling water. The solution is more acidic than ethanoic acid solution. In benzene solution it exists as its dimer $(C_6H_5COOH)_2$ because of hydrogen bonding between molecules.

Chemical properties

1. *Acidity.*

(a) Aqueous solutions are acidic:

$$CH_3COOH + H_2O \rightleftharpoons H_3O^+ + CH_3COO^-$$

ethanoate ion

(R = alkyl)

$$RCOOH + H_2O \rightleftharpoons H_3O^+ + RCOO^-$$

alkanoate ion

(b) Sodium and potassium hydroxides form salts and water:

$$CH_3COOH + NaOH \rightarrow H_2O + CH_3COONa$$

sodium ethanoate

(c) Sodium carbonate and sodium hydrogencarbonate react with carboxylic acids and carbon dioxide is evolved:

$$2CH_3COOH(aq) + CO_3^{2-}(aq) \rightarrow H_2O + CO_2(g) + 2CH_3COO^-(aq)$$

$$CH_3COOH(aq) + HCO_3^-(aq) \rightarrow H_2O + CO_2(g) + CH_3COO^-(aq)$$

Sodium hydrogencarbonate reacts with benzoic acid but has no action on phenol (p. 130). Therefore a mixture of this acid and phenol can be separated by adding it to aqueous sodium hydrogencarbonate, which forms the sodium salt of the acid. The unchanged phenol is extracted with ether and the sodium benzoate remains in aqueous solution. Addition of a mineral acid precipitates benzoic acid:

$$\left\{\begin{matrix} C_6H_5COOH \\ \\ C_6H_5OH \end{matrix}\right. \xrightarrow{NaHCO_3(aq)} \left\{\begin{matrix} C_6H_5COONa \xrightarrow{H_3O^+} C_6H_5COOH(s) \\ \\ C_6H_5OH \xrightarrow{ether} C_6H_5OH \text{ (in ether)} \end{matrix}\right.$$

(d) Sodium, potassium and magnesium liberate hydrogen from the acids:

$$2RCOOH + Mg \rightarrow H_2(g) + (RCOO)_2Mg$$

2. *Alcohols* (formation of esters). An acid and an alcohol react reversibly to form an ester and water; the reaction is catalysed by oxonium ions:

$$RCOOH + HOR' \xrightarrow{H_3O^+} H_2O + RCOOR'$$

$$C_6H_5COOH + HOCH_3 \rightleftharpoons H_2O + C_6H_5COOCH_3$$

3. *Phosphorus chlorides and sulphur dichloride oxide.* Phosphorus pentachloride reacts vigorously and the trichloride reacts less vigorously to form alkanoyl chlorides: steamy fumes of hydrogen chloride are evolved indicating the presence of hydroxyl groups in the acids:

$$CH_3COOH(l) + PCl_5(s) \rightarrow HCl(g) + PCl_3O + CH_3COCl(l)$$

ethanoyl chloride

$$3CH_3COOH(l) + PCl_3(l) \rightarrow H_2PHO_3 + 3CH_3COCl(l)$$

$$\begin{matrix} COOH \\ | \\ COOH \end{matrix} \xrightarrow[SOCl_2]{PCl_5 \text{ or}} \begin{matrix} COCl \\ | \\ COCl \end{matrix}$$

ethanedioyl chloride

$$C_6H_5COOH(s) + SCl_2O(l) \rightarrow HCl(g) + SO_2(g) + C_6H_5COCl(l)$$

benzoyl chloride

In these reactions the OH group of the acids resembles that of alcohols (p. 113). Methanoic acid does not form the compound HCOCl, which does not exist:

$$HCOOH + PCl_5 \rightarrow 2HCl + PCl_3O + CO$$

4. *Halogens.* Chlorine or bromine reacts with boiling acids in sunlight or ultra-violet light, preferably with a catalyst (iodine, sulphur or phosphorus). These catalysts are *halogen carriers.* The hydrogen atoms of the CH_3 group in ethanoic acid are replaced successively, but that of the OH group is not replaced:

$$CH_3COOH + Cl_2 \rightarrow HCl(g) + CH_2ClCOOH$$

chloroethanoic acid

$$CH_2ClCOOH + Cl_2 \rightarrow HCl(g) + CHCl_2COOH$$

dichloroethanoic acid

$$CHCl_2COOH + Cl_2 \rightarrow HCl(g) + CCl_3COOH$$

trichloroethanoic acid

The reaction can be stopped at any stage by noting when the calculated increase in mass has occurred.

5. *Reduction.* Lithium tetrahydridoaluminate(III) in ethereal solution converts the acids to primary alcohols:

$$CH_3COOH \xrightarrow{Li[AlH_4]} CH_3CH_2OH$$

6. *Sodalime.* Hydrocarbons are formed when the acids (or usually their sodium salts) are heated strongly with sodalime or calcium oxide. *Decarboxylation* (loss of CO_2 from the COOH group) occurs:

$$C_6H_5COOH(s) + 2NaOH(s) \rightarrow Na_2CO_3 + H_2O + C_6H_6(g)$$

$$CH_3COONa(s) + NaOH(s) \rightarrow Na_2CO_3 + CH_4(g)$$

7. *Iron(III) chloride.* The chloride produces a deep red colour with neutral salts of methanoic and ethanoic acids. The colour is that of the iron(III) compounds, which on boiling form brown precipitates of basic alkanoates:

$$3CH_3COO^-(aq) + Fe^{3+}(aq) \rightarrow (CH_3COO)_3Fe(aq), \text{ red}$$

$$(CH_3COO)_3Fe(aq) + 2H_2O \rightarrow 2CH_3COOH + CH_3COOFe(OH)_2(s)$$

Special properties of methanoic acid

Each molecule of this acid contains the aldehyde group –CHO. Compare the following:

methanal	methanoic acid	ethanoic acid
$H-\underset{\underset{O}{\parallel}}{C}-H$	$HO-\underset{\underset{O}{\parallel}}{C}-H$	$HO-\underset{\underset{O}{\parallel}}{C}-CH_3$

Methanoic acid is a powerful reducing agent because it is readily oxidized to carbon dioxide and water:

$$HCOOH \xrightarrow{[O]} CO_2 + H_2O \quad \text{or} \quad HCOOH - 2e^- \rightarrow CO_2 + 2H^+$$

1. *Tollen's reagent.* The reagent is reduced to a black precipitate of silver. It is difficult to form a silver mirror (p. 153) because methanoates are such powerful reducing agents, but one is sometimes produced with very dilute solutions:

$$HCOOH(aq) + Ag^+(aq) + 2OH^-(aq) \rightarrow Ag(s) + CO_2(g) + 2H_2O$$

(Methanoic acid has no action on Fehling's solution.)

2. *Potassium manganate(VII) and dichromate(VI).* The pink manganate(VII) solution is decolorized even in the cold, and the yellow or orange dichromate is converted to green chromium(III) salts.

3. *Mercury(II) chloride.* This is reduced to a white precipitate of mercury(I) chloride and then to grey liquid mercury:

$$HCOOH(aq) + 2HgCl_2(aq) \rightarrow Hg_2Cl_2(s) + CO_2 + 2HCl$$

$$HCOOH(aq) + Hg_2Cl_2(s) \rightarrow 2Hg(l) + CO_2 + 2HCl$$

4. *Concentrated sulphuric acid.* The acid and its salts are converted to carbon monoxide. The acid is dehydrated. No charring occurs (usually sulphuric acid chars organic compounds):

$$\begin{matrix} H \diagdown \\ HO \diagup \end{matrix} C{=}O(l) \rightarrow CO(g) + H_2O$$

Special properties of benzoic acid

This acid has a benzene ring in its molecule, and therefore it can be nitrated, sulphonated, chlorinated and brominated. The carboxyl group directs the entering group to the 3-position:

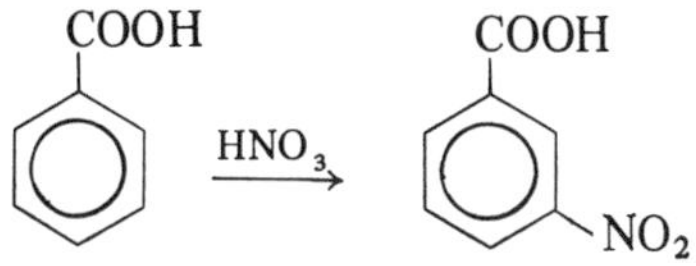

$$C_6H_5COOH \xrightarrow{HNO_3} C_6H_4(NO_2)COOH$$

3-nitrobenzoic acid

Displayed (graphic) formulae of ethanoic acid (acetic acid)

Qualitative and quantitative analysis shows that the empirical formula of the acid is CH_2O. The relative molecular mass when dissolved in water or ethanol is 60; therefore the molecular formula is $C_2H_4O_2$.

The presence of one hydroxyl group is shown by the action of phosphorus pentachloride, which replaces one oxygen and one hydrogen atom by one chlorine atom:

$$C_2H_3O{-}OH \xrightarrow{PCl_5} C_2H_3O{-}Cl$$

This is confirmed by the action of sodium, which replaces only one of the four hydrogen atoms.

The presence of one methyl group is confirmed by synthesis of the acid from methanol as follows:

$$CH_3OH \xrightarrow{P + I_2} CH_3I \xrightarrow{KCN} CH_3CN \xrightarrow{H_3O^+} CH_3COOH$$

The structure of the acid is therefore

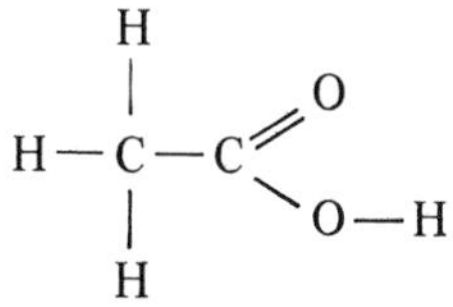

$$H{-}\overset{\overset{H}{|}}{\underset{\underset{H}{|}}{C}}{-}C\begin{matrix}\nearrow\!\!\!\!\!\nearrow O \\ \searrow O{-}H\end{matrix}$$

However, this structure does not explain its acidity. To understand this it is necessary to consider the structure and stability of the carboxylate ion.

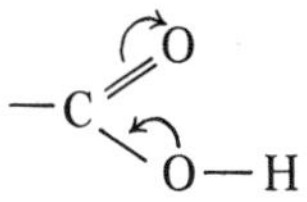

$$-C\begin{matrix}\nearrow\!\!\!\!\!\nearrow \curvearrowright O \\ \searrow \curvearrowleft O{-}H\end{matrix}$$

The partial positive charge on the oxygen atom of the OH group means that the hydrogen atom (as a proton) can leave more readily. The OH group in carboxylic acids is therefore more acidic than the OH group in alcohols, in which delocalization cannot occur.

Carboxylate ion. Alternative formulae are:

$$-C\begin{matrix}\nearrow O \\ \searrow O^-\end{matrix} \qquad\qquad -C\begin{matrix}\nearrow O^- \\ \searrow O\end{matrix}$$

There is no reason why the double bond should be between the carbon atom and one particular oxygen atom. The true structure is intermediate between the above two extreme forms. One electron pair is not localized between two atoms but is spread out or delocalized over the three atoms. Delocalization of electrons is represented by a dotted line, representing a partial bond:

$$\left.-C\begin{matrix} O \\ O\end{matrix}\right\}^- \quad \text{or} \quad \left[-C\begin{matrix} O \\ O\end{matrix}\right]^-$$

The bond lengths and angles in the methanoate ion, $HCOO^-$, have been determined by X-ray diffraction methods. The lengths of the carbon–oxygen bonds are equal and are 0.127 nm, and the angle between the two bonds is 124°. The usual lengths of bonds are: C—O 0.143 nm, and C=O 0.122 nm. The figures show that the carbon–oxygen bonds in the ion are identical and are neither true single nor true double bonds.

The delocalization of the negative charge in the carboxylate ion means that the ion is less ready to accept a hydrogen ion than is the alkoxide ion formed by the ionization of alcohols, i.e. it acts as a weaker base. Therefore the OH group in carboxylic acids is more acidic than that in alcohols.

Uses of carboxylic acids

Vinegar is a dilute solution of ethanoic acid. The acid is used in the manufacture of cellulose ethanoate (used for making lacquers and rayon (p. 173)), paints containing its lead and copper salts, to coagulate rubber latex, and in the manufacture of propanone and ethanoic anhydride. Iron and aluminium ethanoates are used for dyeing cotton. Ethanoic acid is a solvent in reactions during which oxidation occurs because the acid is very difficult to oxidize.

Methanoic acid is used in the dyeing of wood and cotton, to free hides from the calcium oxide used to remove hairs (during the making of leather from hides), as a preservative for some fruit juices, and to coagulate rubber latex.

Benzoic acid is a preservative used in some foodstuffs and soft drinks, and its ammonium salt is used in medicine.

Ethanedioic acid is used in the dyeing industry and in the manufacture of some inks. It is used in laundries to remove iron stains and ink spots on clothing (it reduces insoluble Fe^{3+} compounds to soluble Fe^{2+} compounds).

Strengths of acids

A strong acid is completely dissociated in aqueous solution; a weak acid is slightly dissociated. The relative strengths of weak acids are indicated by their dissociation constants K_a. Consider the weak acid HA:

$$HA + H_2O \rightleftharpoons H_3O^+ + A^-$$

By the equilibrium law, $K_a = \frac{[H_3O^+][A^-]}{[HA]} \text{ mol dm}^{-3}$ (the value of $[H_2O]$ is practically a constant and is omitted from the denominator of the equation).

The relative strengths are also compared by pK_a values, and

$$pK_a = -\lg(K_a/\text{mol dm}^{-3})$$

A strong acid has a low pK_a value.

	$K_a/\text{mol dm}^{-3}$	pK_a
$HCOOH$	2.0×10^{-4}	3.7
CH_3COOH	1.8×10^{-5}	4.75
$CH_2ClCOOH$	1.4×10^{-3}	2.85
$CHCl_2COOH$	5.0×10^{-2}	1.3
CCl_3COOH	1.3×10^{-1}	0.9
CH_2FCOOH	2.0×10^{-3}	2.7
$CH_2ClCOOH$	1.4×10^{-3}	2.85
$CH_2BrCOOH$	1.3×10^{-3}	2.9
CH_2ICOOH	1.0×10^{-3}	3.0
$ClCH_2CH_2COOH$	8.3×10^{-5}	4.1
CH_3CH_2COOH	1.3×10^{-5}	4.9
$CH_3CH_2CH_2COOH$	1.29×10^{-5}	4.9
$CH_3(CH_2)_4COOH$	1.28×10^{-5}	4.9
C_6H_5–COOH	6.3×10^{-5}	4.2
CH_3–C_6H_4–COOH	4.6×10^{-5}	4.34

Influence of substituents on acidity

Effect of chlorine. The strengths of chlorine-substituted acids are in these orders:

$$CCl_3COOH > CHCl_2COOH > CH_2ClCOOH > CH_3COOH$$

$$CH_2ClCOOH > ClCH_2CH_2COOH > CH_3COOH$$

The carbon–chlorine bond is polar, $\overset{+}{C}-\overset{-}{Cl}$ The electronegative chlorine atom is electron-attracting, and it pulls electrons away from the carbon. Therefore the first carbon atom (in $CH_2ClCOOH$) is positive and it pulls harder on the electron pair it shares with the carbon atom of the COOH group. As a result the OH bond is made more polar and the hydrogen atom (as a proton) is more readily detached, i.e. the acid more readily ionizes.

The transmission of electrical effects from atom to atom in a chain (e.g. from Cl to C, then to the second C, then to the OH group) by successive distortion of shared electrons is called an *inductive effect.* The inductive effects of two chlorine atoms in $CHCl_2COOH$ and of three in CCl_3COOH are clearly much greater, and therefore the acids are stronger. An arrow indicates the inductive effect and polarization of a bond; the head points towards the direction in which electrons are pulled:

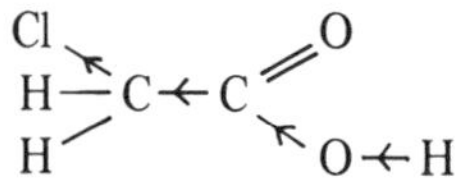

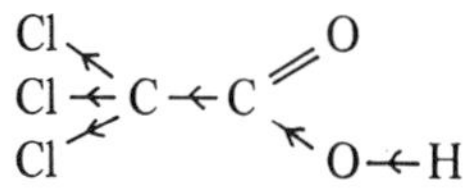

```
Cl↖         O              Cl↖          O
H—C ← C ⫽                 Cl ← C ← C ⫽
H/      ↖ O ← H            Cl↙       ↖ O ← H
```

The acid $ClCH_2CH_2COOH$ is weaker than the acid $ClCH_2COOH$ (but is stronger than $ClCH_2CH_2CH_2COOH$) because the inductive and polarization effects of the chlorine atom are smaller the further away they are from the COOH group:

```
                     Cl ← C ← C ← C ←
                              ↑    ↑    ↑
Polarization is            weak weaker very weak
```

Effects of halogens. Of the four halogen atoms (F, Cl, Br and I), fluorine is most electronegative and iodine is least (p. 93). Therefore the strengths of halogen-substituted acids are in the order:

$$CH_2FCOOH > CH_2ClCOOH > CH_2BrCOOH > CH_2ICOOH$$

Effect of alkyl groups. The acidities of alkanoic acids are in the order:

$$HCOOH > CH_3COOH > CH_3CH_2COOH > CH_3CH_2CH_2COOH$$

and so on.

An alkyl group has an electron-releasing effect (relative to a hydrogen atom). Therefore the effect of an alkyl group is the opposite of that of a chlorine atom, and each additional CH_2 group makes the acid weaker.

Effects of other groups. Benzoic acid is almost twice as strong as ethanoic acid. This is because of the electron-attracting power of the phenyl group, C_6H_5; this power is observed in the properties of phenol and phenylamine.

Consider the dissociation constants below and relate their values to the structures of the acids:

	$K_a/mol\,dm^{-3}$
2-Nitrobenzoic	61×10^{-4}
3-Nitrobenzoic	3.5×10^{-4}
4-Nitrobenzoic	4.0×10^{-4}
2-Hydroxybenzoic acid	10×10^{-4}
3-Hydroxybenzoic acid	0.9×10^{-4}
4-Hydroxybenzoic acid	0.3×10^{-4}

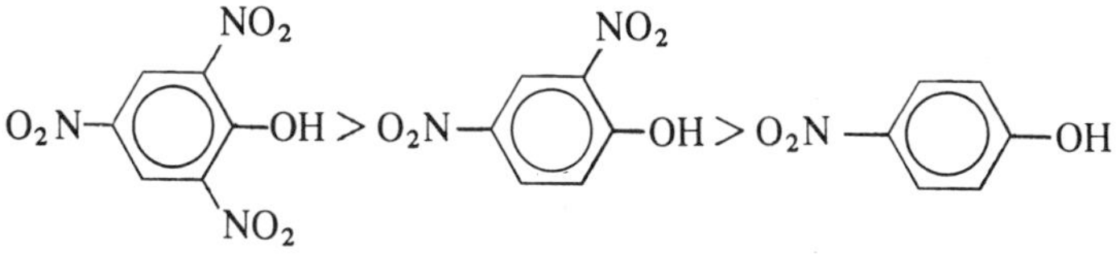

2,4,6-trinitrophenol (O_2N, NO_2, NO_2, OH) > 2,4-dinitrophenol (O_2N, NO_2, OH) > 4-nitrophenol (O_2N, OH)

$K_a = 2 \times 10^{-1}$ $\quad K_a = 1.3 \times 10^{-5}$ $\quad K_a = 6.2 \times 10^{-8}$

SUMMARY

Carboxylic acids

R—COOH Functional group: —COOH or $-\overset{\delta+}{C}(=\overset{\delta-}{O})\leftarrow O \leftarrow H$.

Preparation

1. Oxidize primary alcohols or aldehydes with MnO_4^-/H^+.
2. Reflux nitrile with acid or alkali.
3. Oxidize aromatic hydrocarbons with $HNO_3(aq)$ or MnO_4^-/H^+.

Properties

Reagent/reaction	*Ethanoic acid, CH_3COOH*	*Benzoic acid, C_6H_5COOH*
Acidity		
H_2O	Acidic solution	Acidic (stronger acid)
Alkalis	Salts, ethanoates	Salts, benzoates
HCO_3^- or CO_3^{2-}	CO_2 evolved	CO_2 evolved
Na, K or Mg	H_2 + ethanoates	H_2 + benzoates
C_2H_5OH/H^+, reflux	$CH_3COOC_2H_5$ ethyl ethanoate	$C_6H_5COOC_2H_5$ ethyl benzoate
PCl_5, PCl_3 or SCl_2O	CH_3COCl ethanoyl chloride	C_6H_5COCl benzoyl chloride
$Cl_2(g)/I_2$ or P, boil	$CH_2ClCOOH$, $CHCl_2COOH$, then CCl_3COOH	No similar reaction
$Li[AlH_4]$ (ether)	CH_3CH_2OH ethanol	$C_6H_5CH_2OH$ phenylmethanol
Sodalime, heat	CH_4 methane	C_6H_6 benzene

Benzoic acid can be nitrated, sulphonated and halogenated, forming the 3-derivative, e.g. $C_6H_4(NO_2)COOH$, 3-nitrobenzoic acid.

Methanoic acid HCOOH is also an aldehyde HO—CHO. It reduces (a) Tollen's reagent, $[Ag(NH_3)_2]^+$, to black silver (no silver mirror), (b) $KMnO_4/H^+$ and $K_2Cr_2O_7/H^+$ even when cold, and (c) $HgCl_2(aq)$ to mercury. H_2SO_4(conc) dehydrates to CO(g) without charring. It does not reduce Fehling's solution, unlike other aldehydes.

Manufacture

Ethanoic acid

(a) Petroleum $\xrightarrow{\text{crack}}$ methane $\xrightarrow{\text{air/catalyst}}$ ethanal, then ethanoic acid.

(b) Petroleum $\xrightarrow{\text{crack}}$ pentane, hexane, heptane $\xrightarrow[\text{pressure}]{\text{air/catalyst}}$

mixture of alkanoic acids $\xrightarrow{\text{fractionation}}$ ethanoic acid (and others).

(c) Fermentation of beer and wines → ethanoic acid (vinegar).

Hexane-1,6-dioic acid, $HOOC(CH_2)_4COOH$ (used to make nylon 66).

$$\text{Benzene} \xrightarrow{H_2/Ni} C_6H_{12} \xrightarrow{\text{air/catalyst}} C_6H_{11}OH \xrightarrow{HNO_3} \text{Acid.}$$

cyclohexanol ($C_6H_{11}OH$) is also obtained from Phenol $\xrightarrow{H_2/Ni}$

Relative strengths of acids, RCOOH. Acidity is greater if R is electron-attracting (halogens, C_6H_5, NO_2, OH), weaker if electron-donating (alkyl groups).

$$CCl_3COOH > CHCl_2COOH > CH_2ClCOOH > CH_3COOH$$

$$CH_2FCOOH > CH_2ClCOOH > CH_2BrCOOH > CH_2ICOOH$$

$$HCOOH > CH_3COOH > CH_3CH_2COOH > CH_3CH_2CH_2COOH$$

$$C_6H_5COOH > CH_3COOH > C_6H_5OH > C_2H_5OH$$

QUESTIONS

1. How would you separate and isolate methanoic (formic) acid from its aqueous solution? Explain why methanoic acid gives a red precipitate with Fehling's solution but ethanoic (acetic) acid does not.

2. How does the hydroxyl group in ethanol and in ethanoic (acetic) acid react with (a) water, (b) phosphorus chlorides, and (c) sulphur dichloride oxide?

3. How are the chloroethanoic (chloroacetic) acids prepared? How do the strengths of these acids compare with that of ethanoic (acetic) acid? Account for the differences in strengths of the four acids.

4. Compare and contrast ethanoic acid and benzoic acid with respect to their acidic properties and their actions on phosphorus pentachloride, ethanol and sodalime.

5. Describe simple experiments by which you would distinguish between (a) benzoic acid and phenol, (b) aqueous sodium ethanoate and aqueous sodium benzoate, (c) ethanoic acid and ethanedioic acid.

6. Outline two methods by which benzoic acid may be obtained from benzene and one method by which it is prepared from methylbenzene. In what ways do the reactions of benzoic acid and benzenesulphonic acid differ?

7. A mixture of three organic compounds was treated with excess of dilute sodium hydroxide. Part of it dissolved and the undissolved portion was a colourless liquid which burned with a very smoky flame and when treated with a mixture of nitric and sulphuric acids gave a pale yellow liquid which smelled of almonds. The aqueous solution when treated with carbon dioxide gave an

oily substance which burned with a smoky flame and gave a violet colour with iron(III) chloride. The final aqueous liquid was acidified and distilled; a strongly acid distillate was obtained which was neutralized with sodium hydroxide and evaporated to dryness. The resulting colourless solid, with warm concentrated sulphuric acid, gave carbon monoxide as the gaseous product. Explain the above reactions, identify the three components of the mixture, and give one further chemical test for any two of them. (C.)

8. From your knowledge of the reactions of particular groups predict the reactions of the compound:

$$HO-C_6H_4-CH=CH-CH_2-OH$$

with (a) bromine water, (b) hydrogen and finely divided nickel, (c) sodium hydroxide, (d) ethanoic acid, and (e) dilute alkaline potassium manganate(VII), $KMnO_4$. (C.)

9. Write explanatory notes on the following. (a) The formula of ethanoic acid contains the carbonyl group, but the acid shows none of the addition and condensation reactions characteristic of this group. (b) When ethene is passed through bromine water containing sodium chloride, 1-bromo-2-chloroethane can be detected among the products formed. (c) There is only one compound CH_2Cl_2, but two compounds $C_2H_4Cl_2$ and three compounds $C_2H_2Cl_2$. (d) When phenol is nitrated the product is a mixture of 2- and 4-nitrophenol with hardly any 3-nitrophenol. (L.)

Chapter 14

ACID CHLORIDES AND ACID ANHYDRIDES

Acid chlorides

An acid chloride (or acyl chloride) is a compound in which the OH group of an alkanoic or aromatic carboxylic acid has been replaced by a chlorine atom. The general formula is RCOCl or

$$\overset{\delta+}{R-C}(=O^{\delta-})-Cl$$

in which R is an alkyl or aryl group. Methanoyl chloride, HCOCl, is unknown. Two common acid chlorides are:

CH_3COCl	C_6H_5COCl
$CH_3-C(=O)-Cl$	$C_6H_5-C(=O)-Cl$
ethanoyl chloride (acetyl chloride)	benzoyl chloride, or benzenecarbonyl chloride

The ethanoyl (acetyl) group CH_3CO- is an *acyl* group.

Preparation of acid chlorides

They are prepared by the action on acids of phosphorus pentachloride, at room temperature, or of phosphorus trichloride, which usually requires heating, or of sulphur dichloride oxide. The last compound reacts least readily but the gaseous by-products are easily separated from the acid chlorides.

$$CH_3COOH(l) + PCl_5(s) \rightarrow CH_3COCl(l) + PCl_3O(l) + HCl(g)$$

$$3CH_3COOH(l) + PCl_3(l) \rightarrow 3CH_3COCl(l) + H_2PHO_3$$

$$CH_3COOH(l) + SCl_2O(l) \rightarrow CH_3COCl(l) + SO_2(g) + HCl(g)$$

Ethanoyl chloride distils over when heated at 324 K. PCl_3O is phosphorus trichloride oxide, a colourless fuming liquid (b.p. 380 K) and H_2PHO_3 is phosphonic acid.

Benzoyl chloride (b.p. 470 K) is made by similar reactions. Phosphorus pentachloride reacts at room temperature; the phosphorus trichloride oxide is then distilled off first.

$$C_6H_5COOH + PCl_5 \rightarrow C_6H_5COCl + PCl_3O + HCl$$

Acid anhydrides

An acid anhydride may be regarded as a compound produced (theoretically and not in practice) by removal of one molecule of water from two molecules of carboxylic acid:

$$RCOOH + HOOCR \rightarrow (RCO)_2O + H_2O \quad (R = \text{alkyl or aryl group})$$

$$\begin{matrix} R-C\!=\!O \\ \;\;\;\;\;\;\;\;\; \backslash O \\ R-C\!=\!O \end{matrix} \quad \text{e.g.} \quad \begin{matrix} CH_3-C\!=\!O \\ \;\;\;\;\;\;\;\;\;\;\;\;\; \backslash O \\ CH_3-C\!=\!O \end{matrix} \quad \text{or} \quad (CH_3CO)_2O$$

ethanoic anhydride (acetic anhydride)

$(C_6H_5CO)_2O$ is benzoic anhydride. $(HCO)_2O$ or $H_2C_2O_3$, methanoic anhydride, does not exist.

Preparation of acid anhydrides

Add an acid chloride to an anhydrous sodium salt of the corresponding carboxylic acid and distil off the anhydride:

$$CH_3COCl(l) + NaOOCCH_3(s) \rightarrow NaCl + (CH_3CO)_2O(l)$$

$$C_6H_5COCl(l) + NaOOCC_6H_5(s) \rightarrow NaCl + (C_6H_5CO)_2O(l)$$

Reactions of acid chlorides and acid anhydrides

(Reactions of ethanoyl chloride are vigorous and may be dangerous. They should be demonstrated by a teacher.)

1. *Water.* Add about 5 cm^3 of water to a beaker on an asbestos square. Use a teat pipette or glass tube to add a few drops of ethanoyl chloride to the water. Observe the reaction. Divide the product into two portions. Test one portion for chloride ions by adding aqueous silver nitrate. Test the second portion for ethanoic acid by neutralizing it with dilute ammonia solution and then adding iron(III) chloride solution (p. 179).

Repeat the test with benzoyl chloride; warm if necessary.

Repeat the test with ethanoic anhydride. Note if it mixes with cold water; warm if necessary.

2. *Ethanol.* Add about 5 cm^3 of ethanol to a beaker. Use a teat pipette to add a few drops of ethanoyl chloride. Shake for two minutes. When reaction is finished, add aqueous sodium carbonate until no more carbon dioxide is evolved. Identify the product by its smell.

Repeat the test with benzoyl chloride and then with ethanoic anhydride, and warm if necessary.

3. *Phenol.* Refer to test 7 on p. 129.

4. *Ammonia.* Add about 5 cm^3 of concentrated ammonia solution to a beaker. Add carefully a few drops of ethanoyl chloride.

Repeat the test with benzoyl chloride and then with ethanoic anhydride; warm if necessary.

5. *Phenylamine (aniline).*
(a) Add a few drops of phenylamine to a beaker. Add ethanoyl chloride, one drop at a time. (Be prepared for a vigorous reaction.) When reaction is finished, add about 40 cm^3 of water, boil, and allow the mixture to cool and crystallize.

(b) Mix 1 cm^3 of phenylamine with 4 cm^3 of dilute sodium hydroxide in a beaker and warm. Add 2 cm^3 of benzoyl chloride and warm gently for about 10 minutes. Pour the product into cold water. Filter off the precipitate. Recrystallize from ethanol if necessary.

(c) Warm a mixture of phenylamine and ethanoic anhydride (a few drops of each), and then pour the product into about 10 cm^3 of water.

Properties of acid chlorides and acid anhydrides

Physical properties

Ethanoyl chloride is a colourless volatile liquid with a pungent irritating smell. In moist air it forms fumes of hydrogen chloride and ethanoic acid; therefore it is stored in bottles with ground-glass stoppers sealed into the necks with wax.

Benzoyl chloride (b.p. 471 K) is a colourless fuming liquid with a penetrating and irritating smell. It is denser than and almost insoluble in water.

Ethanoic anhydride (b.p. 413 K) is a colourless liquid with a sharp irritating smell. It does not fume in moist air. It is denser than and not very soluble in cold water.

Benzoic anhydride is a white crystalline solid (m.p. 315 K, b.p. 633 K). It is insoluble in water.

Chemical properties. Ethanoyl chloride is one of the most reactive compounds; ethanoic anhydride and benzoyl chloride are much less reactive. In the —COCl group, the strongly electronegative chlorine atom increases the polarity of the carbonyl group:

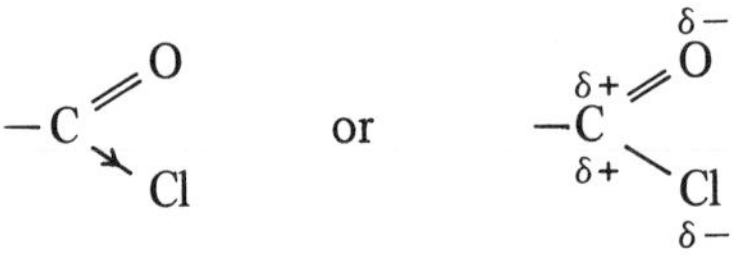

—C(=O)(Cl) or —C$^{\delta+}$(=O$^{\delta-}$)(Cl$^{\delta-}$)

The carbon atom is so positive that even weak nucleophiles such as water and alcohols can react with it. Benzoyl chloride is less reactive because the positive charge is delocalized over the benzene ring:

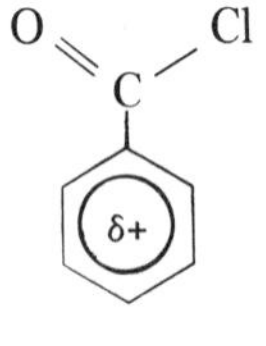

C_6H_5COCl (ring $\delta+$)

1. *Hydrolysis.* Ethanoyl chloride reacts violently with water (and dangerously with alkalis). Benzoyl chloride reacts very slowly with cold water but rapidly with hot water, and the acid formed crystallizes when the solution is cooled. Ethanoic anhydride has little action on cold water, but reacts with warm water and is readily hydrolysed by alkalis.

$$CH_3COCl + H_2O \rightarrow HCl(g) + CH_3COOH(l \text{ or } aq)$$

$$(CH_3CO)_2O + H_2O \rightarrow 2CH_3COOH(l \text{ or } aq)$$

$$(CH_3CO)_2O + 2NaOH(aq) \rightarrow H_2O + 2CH_3COONa(aq)$$

$$C_6H_5COCl + H_2O \rightarrow HCl(g) + C_6H_5COOH(s)$$

2. *Alcohols and phenols* (replacement of —Cl by —OR.) Acid chlorides react vigorously with alcohols to form esters. Benzoyl chloride also reacts readily with phenol, in alkaline solution.

$$RCOCl + HOR' \rightarrow HCl + RCOOR'$$

$$CH_3COCl + HOC_2H_5 \rightarrow HCl + \underset{\text{ethyl ethanoate}}{CH_3COOC_2H_5}$$

$$C_6H_5COCl + HOC_6H_5 \rightarrow HCl + \underset{\text{phenyl benzoate}}{C_6H_5COOC_6H_5}$$

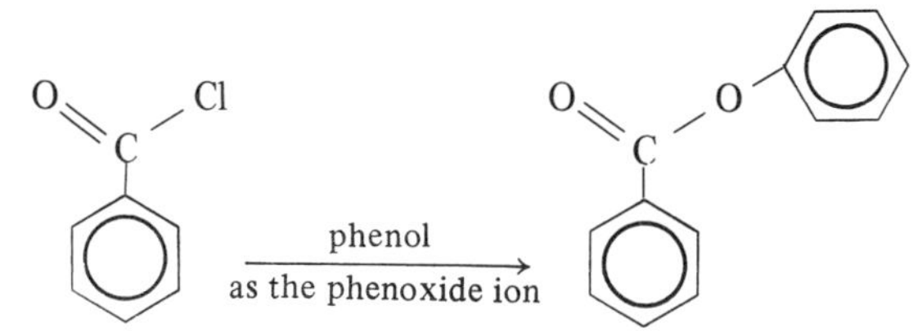

$$C_6H_5COCl \xrightarrow[\text{as the phenoxide ion}]{\text{phenol}} C_6H_5COOC_6H_5$$

benzoyl chloride — phenyl benzoate

Acid anhydrides react less vigorously with alcohols and phenols (in alkaline solution):

$$(CH_3CO)_2O + HOC_2H_5 \rightarrow CH_3COOC_2H_5 + CH_3COOH$$

$$(CH_3CO)_2O + NaOC_6H_5 \rightarrow CH_3COOC_6H_5 + CH_3COONa$$

3. *Ammonia* (replacement of —Cl by —NH_2). Acid chlorides and anhydrides react violently with cold concentrated ammonia solution, and amides are formed.

$$CH_3COCl + 2NH_3 \rightarrow NH_4Cl + \underset{\text{ethanamide}}{CH_3CONH_2}$$

$$(CH_3CO)_2O + 2NH_3 \rightarrow CH_3COONH_4 + \underset{\text{ethanamide}}{CH_3CONH_2}$$

$$C_6H_5COCl + 2NH_3 \rightarrow NH_4Cl + \underset{\text{benzamide}}{C_6H_5CONH_2}$$

The reaction with acid chlorides may be represented by:

$$RCOCl + H{-}NH_2 \rightarrow HCl + RCONH_2$$

The hydrochloric acid reacts with more ammonia to form ammonium chloride.

4. *Amines.* Primary amines react vigorously with acid chlorides and anhydrides:

$$CH_3COCl + H_2NC_2H_5 \rightarrow HCl + \underset{N\text{-ethylethanamide}}{CH_3CONHC_2H_5}$$

$$C_6H_5COCl + H_2NC_6H_5 \rightarrow HCl + \underset{N\text{-phenylbenzamide}}{C_6H_5CONHC_6H_5}$$

$$(CH_3CO)_2O + H_2NC_2H_5 \rightarrow CH_3COOH + CH_3CONHC_2H_5$$

N-ethylethanamide

$$CH_3COCl + H_2NC_6H_5 \rightarrow HCl + CH_3CONHC_6H_5$$

N-phenylethanamide

The *N* in the above names means that the ethyl and phenyl groups are attached to a nitrogen atom and not a carbon atom.

5. *Friedel-Crafts reaction.* Acid chlorides and benzene form ketones (p. 79).

6. *Reduction.* Acid chlorides can be reduced to aldehydes by hydrogen in the presence of a palladium catalyst (if the catalyst is made too effective the reduction produces alcohols):

$$RCOCl + H_2 \rightarrow HCl + RCHO$$

Ethanoylation and benzoylation

Ethanoylation (*acetylation*) means the replacement of the H atom of an OH, NH_2 or NHR group by the ethanoyl (acetyl) group CH_3CO. It occurs when ethanoyl chloride or ethanoic anhydride react with alcohols, sugars, phenols, primary amines and secondary amines. The process is used to identify hydroxy- and amino-compounds, especially aromatic compounds, because the products are crystalline, easy to obtain in a pure state, and have definite melting points (compare the use of 2,4-dinitrophenylhydrazine (p. 150)).

In *benzoylation* the C_6H_5CO group replaces a hydrogen atom. It is better than ethanoylation because (a) ethanoyl chloride reacts readily with water whereas benzoyl chloride does not, (b) benzoylation is helped by alkali which both removes acid produced and decomposes excess benzoyl chloride, and (c) the C_6H_5CO derivatives are less soluble than the corresponding CH_3CO derivatives and have higher melting points.

$$C_6H_5COCl + HY \rightarrow C_6H_5COY + HCl$$

$$(Y = C_2H_5O-,\quad C_6H_5O-,\quad C_2H_5NH-,\quad C_6H_5NH- \quad \text{and so on.})$$

$$C_6H_5COCl + 2NaOH \rightarrow C_6H_5COONa + NaCl + H_2O$$

Manufacture of aspirin

2-Hydroxybenzoic acid, HOC_6H_4COOH, reduces pain and fever but irritates the stomach lining. Its ethanoyl ester, $CH_3COOC_6H_4COOH$, is less irritating, and is called aspirin. It is manufactured by the action of carbon dioxide on sodium phenoxide (phenol + sodium hydroxide), then acid to form 2-hydroxy-benzoic acid, followed by ethanoylation with ethanoic anhydride:

$$C_6H_5OH + NaOH + CO_2 \xrightarrow[\text{pressure}]{\text{heat}} \xrightarrow{\text{HCl}} HOC_6H_4COOH$$

$$HOC_6H_4COOH + (CH_3COO)_2O \rightarrow \underset{\text{aspirin}}{CH_3COOC_6H_4COOH}$$

Cellulose ethanoate fibres

Cotton and linen, which are mainly cellulose, can be spun into thread and then made into cloth. Wood cellulose (wood pulp) is also cellulose but its molecules are tangled and cannot be made directly into thread and cloth. Cellulose is a polymer but its repeat unit contains three hydroxyl groups. Either two or three of these $-OH$ groups are ethanoylated to $-OCOCH_3$ by the action of ethanoic anhydride and glacial ethanoic acid. The product is cellulose ethanoate, which is soluble in propanone and other organic solvents. A solution is pumped through a metal disc containing many small holes and the solvent evaporated in a jet of air, leaving threads called acetate (ethanoate) rayon. It is used to make clothing, furnishing fabrics and artificial silk.

SUMMARY

Acid chlorides

RCOCl (R = alkyl or aryl group). Functional group: $-COCl$.

Preparation of acid chlorides
PCl_5 or SCl_2O + acid, room temperature.

Acid anhydrides

$(RCO)_2O$. Functional group: $O{=}C{-}O{-}C{=}O$.

Preparation of acid anhydrides
Acid chloride + anhydrous sodium salt, distil.

Chemical properties
Nucleophiles attack the $C^{\delta+}$ atom. Acid chloride is most reactive. Benzoyl chloride is less reactive than ethanoyl chloride because delocalization over the benzene ring reduces the charge on the C atom of COCl.

Reaction/reagent	*Ethanoyl chloride* CH_3COCl	*Benzoyl chloride* C_6H_5COCl	*Ethanoic anhydride* $(CH_3CO)_2O$
Water forms acids	CH_3COOH Vigorous reaction	C_6H_5COOH Slow. Rapid when hot	CH_3COOH Very slow. More quickly when warm
Alkalis	All form salts, e.g. CH_3COONa		
Alcohols form esters	$CH_3COOC_2H_5$ Vigorous reaction	$C_6H_5COOC_2H_5$ Vigorous reaction	$CH_3COOC_2H_5$ Less vigorous
Phenols/OH^- form esters	$CH_3COOC_6H_5$	$C_6H_5COOC_6H_5$	$CH_3COOC_6H_5$
NH_3(aq) forms amides	CH_3CONH_2 ethanamide	$C_6H_5CONH_2$ benzamide	CH_3CONH_2 ethanamide

Ethanoylation and benzoylation

Replacement of H atom of $-OH$, $-NH_2$ and $=NH$ groups by CH_3CO- or C_6H_5CO-
Used with alcohols, phenols, sugars, primary and secondary amines.

QUESTIONS

1. Outline methods of preparation of ethanoyl chloride from ethanoic acid. How, and under what conditions, does ethanoyl chloride react with (a) water, (b) ethanol, (c) phenol, and (d) ammonia. Write equations for all reactions.

2. Compare and contrast the actions of ethanoyl chloride and of benzoyl chloride on water, ethanol, phenol and ammonia. Account briefly for any differences.

3. State what you would observe in the following reactions and explain the observations: (a) benzoyl chloride is heated with water for some minutes and the product is allowed to cool slowly to room temperature, (b) benzoyl chloride is mixed with sodium phenoxide and sodium hydroxide solution and shaken for some minutes, (c) ethanoic anhydride and concentrated aqueous ammonia are mixed and warmed gently for some time.

4. Give three examples of ethanoylation (acetylation) and three examples (using different reagents) of benzoylation. Explain why these two processes are useful in organic chemistry.

5. Compare and contrast the reactivities of the halogen atoms in chloroethane, ethanoyl chloride, chlorobenzene and tetrachloromethane. Suggest a simple method for estimating chlorine in each of the first two compounds. (C.)

6. Describe one simple chemical test which would enable you to differentiate between the pairs of compounds named below. Mention in each case (i) the reagent(s) you would use, (ii) the conditions of reaction, (iii) the observations made, and (iv) the equation(s) for any reaction(s). The pairs of compounds are (a) ethanoyl chloride and chloroethanoic acid, (b) propanone and ethanal, (c) propane and propene.

7. Each of five reagent bottles, from which the labels have been removed, is known to contain a different one of the following liquids: (a) HCOOH, (b) CH_3COOH, (c) CH_3COCl, (d) $CH_3CH_2CH_2I$, (e) $CH_3CH_2CH_2Cl$. Describe how you would identify each liquid by means of chemical tests that can be readily carried out in a school laboratory. You should name the reagents, and give the essential reaction conditions, the observations made and the conclusions drawn from the observations.

Note: A chemical test is required for each liquid – it is not sufficient to identify the fifth liquid by elimination. (C.)

Chapter 15
GEOMETRICAL AND OPTICAL ISOMERISM

GEOMETRICAL ISOMERISM

Isomers are compounds with the same molecular formula. Isomers whose atoms are bonded together in different sequences are called *structural isomers*, e.g. CH_3CH_2OH and CH_3OCH_3, $CH_3CH_2CH_2CH_3$ and $(CH_3)_2CHCH_3$, 1,2-dinitrobenzene and 1,3-dinitrobenzene, $C_6H_4(NO_2)_2$, and so on. Ethanol and methoxyethane have different functional groups, butane and 2-methylpropane have different carbon skeletons, and dinitrobenzenes have the same functional groups at different places on a benzene nucleus.

Isomers with the same structures exist. For example, there are two butenedioic acids, $C_4H_4O_4$ or $C_2H_2(COOH)_2$ or $HOOC-CH=CH-COOH$. The atoms are arranged differently in space, that is, they have different configurations. Isomers with the same structure but different configurations are called *stereoisomers*, and the phenomenon is *stereoisomerism.* There are two kinds of stereoisomerism – *geometrical isomerism* and *optical isomerism.*

The butenedioic acids are geometrical isomers because the double bond prevents free rotation of the groups in the molecules. The two acids are:

```
H       COOH              HOOC       H
  \    /                      \    /
    C                           C
    ||                          ||
    C                           C
  /    \                      /    \
H       COOH                H       COOH
```

cis-butenedioic acid

(carboxyl groups and hydrogen atoms on *same* side of double bond)

trans-butenedioic acid

(carboxyl groups and hydrogen atoms on *opposite* sides of double bond)

The physical and chemical properties of the butenedioic acids differ:

	Cis	*Trans*
M.p./K	403	560
Solubility/g per 100 g water	79	0.7
Heat of combustion/kJ mol^{-1}	1370	1340

The *cis* acid is more reactive. By heating it just above its melting point it forms an anhydride, which is readily hydrolysed by cold water. These reactions are possible because the two COOH groups are close together. No anhydride of the *trans* acid is known, presumably because the two COOH groups are far apart and cannot form a cyclic anhydride. By heating the *trans* acid at 570 K the *cis* anhydride is formed because the COOH groups first rotate about the double bond:

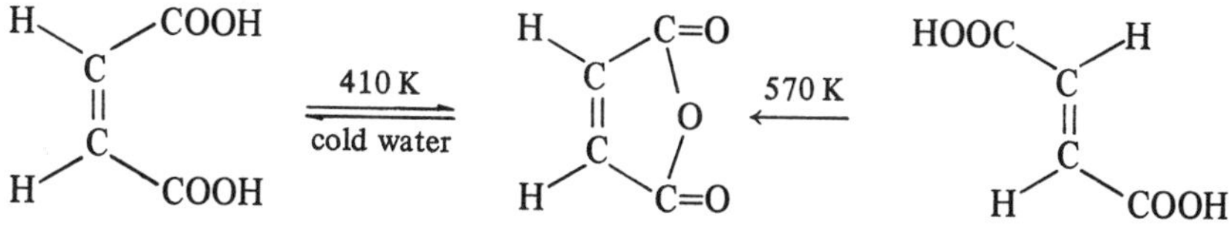

H(COOH)C=C(COOH)H (cis) ⇌ [410 K / cold water] cis anhydride (H–C=C–H with C=O–O–C=O ring) ← [570 K] HOOC(H)C=C(H)COOH (trans)

Both acids form butanedioic acid $HOOCCH_2CH_2COOH$ when reduced.

Geometrical isomers of some oximes exist:

C_6H_5(H)C=N–OH (OH on the side of H), m.p. 308 K

C_6H_5(H)C=N–OH (HO on the side of C_6H_5), m.p.398 K

OPTICAL ISOMERISM

Polarization of light

Ordinary white light consists of waves of many wavelengths which vibrate in all directions at right angles to the path along which the light is travelling. A sodium lamp emits monochromatic light, which is light of one wavelength only, but the vibrations are in all directions. If monochromatic light is passed through Polaroid (a polarizer), only waves vibrating in one particular plane are transmitted and the light is called *plane polarized light.*

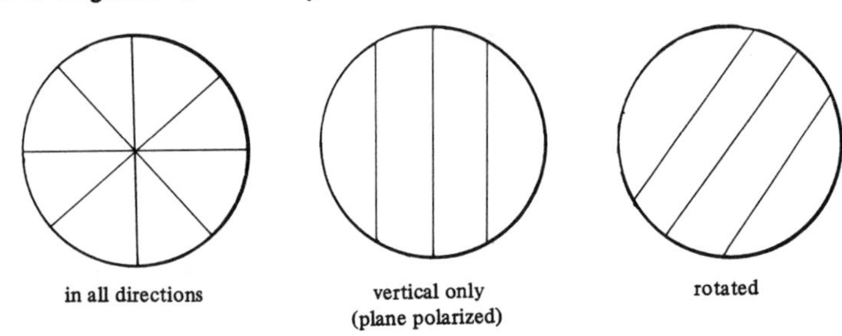

If plane polarized light is passed through a second polarizer, the amount transmitted depends on the angle between the plane of vibration of the light and the plane of vibration of light that the second polarizer can transmit. If the planes are at right angles, no light is transmitted (Fig. 15.1). If the planes are parallel, all the light is transmitted; if the second polarizer is now rotated slowly through 90° the quantity of light gradually diminishes and finally no light is transmitted.

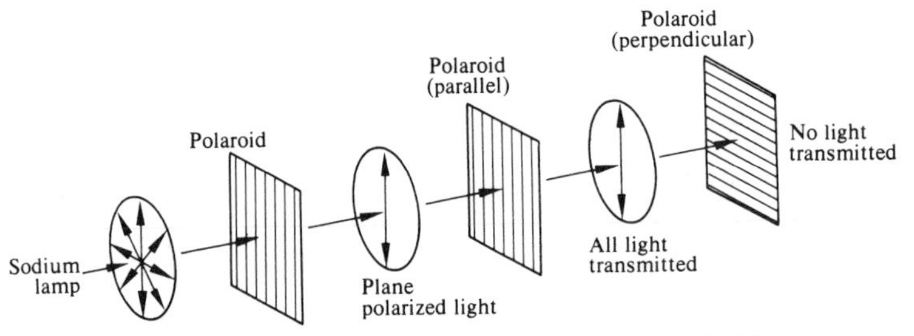

Fig. 15.1. Action of Polaroid on monochromatic light

The instrument used to measure polarization is a *polarimeter*, Fig. 15.2. It consists of:

a sodium lamp to produce monochromatic light,
a fixed polarizer to produce plane polarized light,
a tube 10 to 20 cm long, filled with a solution of the substance being tested, which rotates the plane of polarization,
a second or rotating polarizer, called an analyser, and
a scale to measure the angle of rotation of the light.

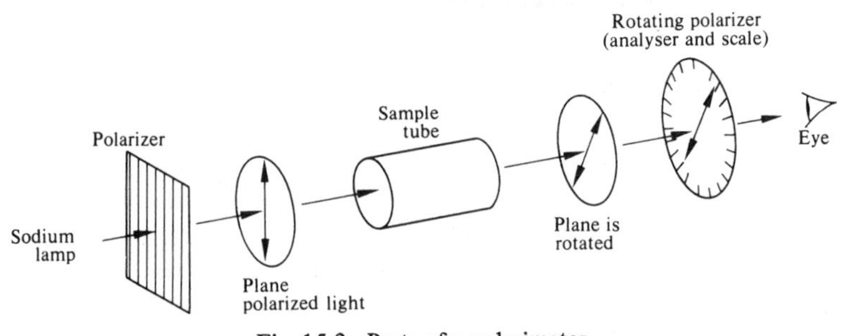

Fig. 15.2. Parts of a polarimeter

Optical rotatory power

Some substances can rotate the plane of polarization of plane polarized light and are *optically active.* The *angle of optical rotation*, which varies for different substances, is the rotation produced by a column of liquid 10 cm long and with a concentration of 1 g of solute in 1 cm^3 of solvent. This angle at 293 K is called the *specific optical rotatory power.*

Rotation of the plane of polarization to the right or in a clockwise direction (as viewed by an observer looking towards the source of light) is given a + sign; a substance having this property is called *dextrorotatory* and has the prefix (+)- to its name, e.g. (+)-lactic acid is an optically active acid obtained from muscle. A substance which rotates the plane of polarization to the left is called *laevorotatory* and has the prefix (−)- to its name, e.g. (−)-lactic acid is obtained from milk. Lactic acid made by synthesis in the laboratory is a mixture of (+) and (−) acids in equal proportions and has the prefix (±). The mixture, which is called a *racemic mixture* or *racemate*, is optically inactive. (The systematic name of lactic acid is 2-hydroxypropanoic acid.)

Optical activity and asymmetry

A symmetrical object (a ball, a cube) can be divided into two identical halves. An asymmetric object (a hand, a foot, a glove) cannot be divided by a plane into two identical halves. One property of an asymmetrical substance is that it cannot be made to coincide with its image in a plane mirror, e.g. the image of a left hand is a right hand. It is impossible to put a left-hand glove on a right hand, and this demonstrates that a left hand and a right hand are not superimposable.

A stereoisomer which, in solution, is optically active is an *optical isomer.* This optical activity depends on the structure of the molecules. In 1848 Pasteur suggested that optically active organic substances were asymmetric; later it was observed that their molecules have four different groups joined to a carbon atom – their formula is Cabcd. A *chiral carbon atom* is one which is bonded to four different atoms or groups. A chiral carbon atom causes an asymmetric molecule, and a compound consisting of such molecules is optically active.

Two optical isomers whose molecules are mirror images of each other are called *enantiomers.* Their structures and geometry are identical, and therefore all ordinary physical and chemical properties are identical. The two isomers rotate the plane of plane polarized light in opposite directions.

Examples of optical isomers

Butan-2-ol and 2-hydroxypropanoic acid (lactic acid) are two compounds containing a chiral carbon atom:

$$\begin{array}{c} H \\ | \\ CH_3-C-OH \\ | \\ CH_2CH_3 \end{array} \qquad\qquad \begin{array}{c} H \\ | \\ CH_3-C-OH \\ | \\ COOH \end{array}$$

butan-2-ol 2-hydroxypropanoic acid

Make models of these molecules, using four different coloured balls to represent the groups attached to the carbon atom. Each can form two different molecules because the valency bonds of the carbon are directed towards the corners of a tetrahedron. The two molecules are mirror images of each other and cannot be superimposed (Fig. 15.3).

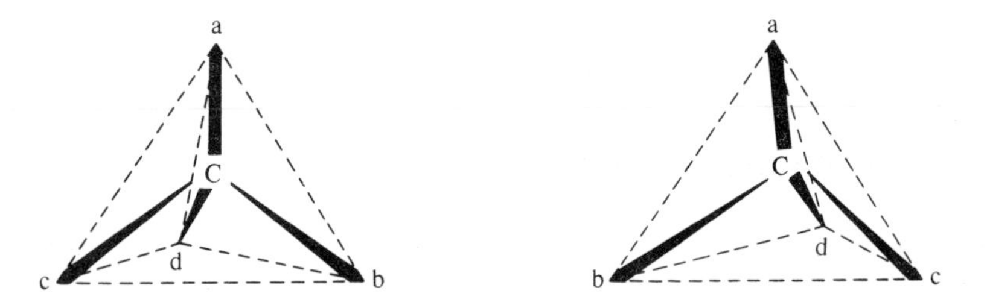

Fig. 15.3. Configurations of asymmetric molecules, Cabcd

A glucose molecule contains four chiral carbon atoms, and sixteen optical isomers have been obtained. Naturally occurring glucose is dextrorotatory. A fructose molecule has three chiral carbon atoms, and the compound is laevorotatory.

Glucose and fructose are formed in equimolar proportions during the hydrolysis of sucrose. Sucrose is dextrorotatory. The rotatory power of glucose is less than that of fructose, and therefore the mixture formed by hydrolysis is laevorotatory. During the hydrolysis of sucrose, the solution is at first dextrorotatory. As the reaction progresses the solution rotates polarized light less and less to the right and gradually becomes optically inactive, and then it becomes laevorotatory. The rate and extent of the hydrolysis can be measured by a polarimeter. The process is called *inversion of sucrose*, and the mixture of (+) glucose and (−) fructose is *invert sugar*.

SUMMARY

Geometrical isomerism

There is restricted rotation about the double bond in alkene molecules. Stereo-isomers have the same structural formula but different arrangements of the bonds in space:

cis

```
X       X
 \     /
  C=C
 /     \
H       H
```

trans

```
H       X
 \     /
  C=C
 /     \
X       H
```

The *trans* isomer is more stable, e.g. higher m.p. *Cis*-butenedioic acid forms an anhydride but the *trans*- acid does not.

Optical isomerism

Optical isomers have the same structural formula but different arrangements of the bonds in space, and rotate the plane of plane polarized light differently. They contain one or more chiral carbon atoms. A chiral carbon atom is attached to four different atoms or groups, e.g. as in butan-2-ol, $CH_3CHOHC_2H_5$, and 2-hydroxypropanoic acid (lactic acid), $CH_3CHOHCOOH$. Two mirror-image isomers (enantiomers) which are not superimposable spatially exist (compare right and left hands). A dextrorotatory or (+)-isomer turns the plane of polarized light to the right, and its laevorotatory (−) enantiomer turns it by an equal amount to the left. An equimolar mixture of enantiomers is a racemic (±) mixture or racemate.

QUESTIONS

1. Explain the meanings of the following terms: polarized light; optically active compound; chiral carbon atom. Name and write the formula of one optically active alcohol and one optically active acid.

2. Discuss the geometrical isomerism of the two butenedioic acids. Explain why one of these acids forms an anhydride but the other does not.

3. Outline briefly the principle of the polarimeter and explain its uses. 2-hydroxypropanoic acid, fructose and glucose contain 1, 3 and 4 chiral carbon atoms per molecule respectively. How many stereoisomers of each compound are possible?

4. A 0.5 M solution of sucrose was contained in a tube 15 cm long and was found to rotate the plane of polarized light through 17° in a clockwise direction. 50 cm^3 of the same solution was then boiled under reflux with 50 cm^3 of 0.5 M sulphuric acid and a sample of the resulting solution was then put in the original tube and the rotation of the polarized light again measured. The reaction is:

$$\text{Sucrose (aq)} + \text{Water} \xrightarrow{H^+(aq)} \text{Glucose} + \text{Fructose}$$

The optical rotatory power of glucose and fructose are +52° and −92° respectively. (a) Draw a diagram showing an arrangement of the apparatus for measuring the rotation of the plane of polarized light. (b) Calculate the optical rotatory power of sucrose. (c) Calculate the reading for the rotation that was observed when the hydrolysed sample was examined. (d) Suggest how the experiment could be modified to determine the dependence of the rate of hydrolysis of sucrose at 293 K on the concentration of hydrogen ions in the solution. (L.)

5. Describe, with diagrams, the type of isomerism shown by trans- and cis-butenedioic acids (fumaric and maleic acids), $(CHCOOH)_2$. Give examples of the differences in physical properties of these acids. How can these acids be distinguished by chemical methods? The addition of hydrogen bromide to each of the two acids gives the same pair of isomers, while the addition of bromine to the two acids gives a total of three isomeric dibromo-acids. Draw the structures of the five brominated compounds and describe the type of isomerism involved. How might the two monobromodicarboxylic acids be separated? (C.)

6. The formula of an acid is $CH_3CHOHCOOH$. What is its systematic name? Outline, by means of equations and essential experimental conditions, how you would prepare this acid from ethanal. Describe an experiment to show that this acid can react as a secondary alcohol. Explain why a solution of this acid obtained from a natural source rotated polarized light to the right, whereas a solution of the acid prepared from ethanal did not rotate polarized light.

7. What do you understand by the terms (a) structural isomerism, (b) ***cis-trans*** isomerism? A gaseous hydrocarbon ***X*** contains 85.7% of carbon by mass. When 0.140 g of ***X*** was introduced into a gas syringe, its volume (after correction to s.t.p.) was found to be 56.0 cm^3. When ***X*** was shaken with aqueous bromine the latter was decolorized. Three structural isomers ***A, B*** and ***C*** of the hydrocarbon ***X*** were found to have the following properties. (i) ***A*** exists as a pair of ***cis-trans*** isomers. (ii) ***B*** underwent oxidation under certain conditions to produce methanal and a compound ***Y*** (empirical formula C_3H_6O) which gave an orange precipitate with 2,4-dinitrophenylhydrazine reagent and a red-brown precipitate on boiling with Fehling's solution. (iii) ***C*** can be obtained by the dehydration of 2-methyl-propan-2-ol. Deduce the full structural formula of each of the isomers ***A, B*** and ***C***, explaining your reasoning and giving balanced equations where possible. (C.)

8. (a) Addition of hydrogen bromide to either of the geometric isomers of but-2-ene leads to a pair of optical isomers. (i) Give the full structural formulae and names of the geometric isomers of but-2-ene. (ii) Draw the full structural formulae of the two optical isomers. (b) Draw a reaction scheme, labelled with reagents and essential conditions to show how but-2-ene may be converted into (i) butanone, (ii) 2-aminobutane. (C.)

Chapter 16
ESTERS

Esters contain the functional group $-COOR$ or $-C\begin{matrix}\nearrow O \\ \searrow OR\end{matrix}$ in which R is an alkyl or aryl group. The general formula is R′COOR, in which R′ is H or an alkyl or aryl group. Examples are:

Methyl methanoate (formate)	$HCOOCH_3$
Ethyl ethanoate (acetate)	$CH_3COOC_2H_5$
Phenyl ethanoate	$CH_3COOC_6H_5$
Ethyl benzoate	$C_6H_5COOC_2H_5$
Phenyl benzoate	$C_6H_5COOC_6H_5$

Isomerism. Esters may be isomeric with each other and with acids:

$HCOOCH_2CH_3$	CH_3COOCH_3	CH_3CH_2COOH
ethyl methanoate	methyl ethanoate	propanoic acid

Five esters are isomers with the formula $C_5H_{10}O_2$:

$CH_3COOCH_2CH_2CH_3$	propyl ethanoate
$CH_3COOCH(CH_3)_2$	1-methylethyl ethanoate
$CH_3CH_2COOCH_2CH_3$	ethyl propanoate
$CH_3CH_2CH_2COOCH_3$	methyl butanoate
$(CH_3)_2CHCOOCH_3$	methyl 2-methylpropanoate

Types of structural isomerism

Chain isomerism. The isomers have different carbon chains or skeletons and are homologues. Examples: butane and methylpropane, $CH_3CH_2CH_2CH_3$ and $(CH_3)_2CHCH_3$.

Position isomerism. The functional group or multiple bond is in different positions in the same carbon skeleton. Examples: propan-1-ol, $CH_3CH_2CH_2OH$, and propan-2-ol, $CH_3CHOHCH_3$; ethoxyethane, $C_2H_5OC_2H_5$, and 2-methoxypropane, $CH_3OCH(CH_3)_2$; 1,2-, 1,3- and 1,4-dichlorobenzenes, $C_6H_4Cl_2$; but-1-ene, $CH_3CH_2CH{=}CH_2$, and but-2-ene, $CH_3CH{=}CHCH_3$.

Functional group isomerism. The isomers have different functional groups and therefore are members of different homologous series. Examples: ethanol, CH_3CH_2OH, and methoxymethane, CH_3OCH_3; propanal CH_3CH_2CHO, and propanone, CH_3COCH_3; ethanoic acid, CH_3COOH, and methyl methanoate, $HCOOCH_3$.

Geometrical isomerism and optical isomerism are two types of stereoisomerism, p. 202, which have different arrangements in space of the same atoms and groups.

Preparation of esters

1. *From an acid and an alcohol.* A carboxylic acid reacts with an alcohol (but not with a phenol). The reaction is slow if the mixture is cold and is not fast even if it is hot. After a long time an equilibrium mixture of an ester and water are formed:

$$R'COOH(l) + HOR(l) \underset{\text{hydrolysis}}{\overset{\text{esterification}}{\rightleftharpoons}} R'COOR(l) + H_2O(l)$$

The mechanism of the reaction (p. 116) involves protonated acid.

Esterification is sometimes compared with neutralization:

$$\text{acid} + \text{alcohol} \rightleftharpoons \text{ester} + \text{water}$$

$$\text{acid} + \text{base} \rightarrow \text{salt} + \text{water}$$

However, esterification is usually very slow and reversible, whereas neutralization is usually very fast (almost instantaneous) and irreversible. Salts and esters are quite different. Salts are ionic solids and can be electrolysed; esters are usually covalent liquids that are only slightly soluble in water.

Fischer-Speier method of esterification. Hydrogen chloride gas is passed through a mixture of alcohol and carboxylic acid heated in a flask fitted with a reflux condenser. (The advantages of hydrogen chloride as a catalyst are that it does not cause formation of alkenes and it does not attack benzene compounds; concentrated sulphuric acid causes both of these side-reactions and they reduce the yield of ester.)

2. *From an acid chloride or acid anhydride and an alcohol or a phenol.* Acid chlorides react readily even in the cold with alcohols and with phenols in alkaline solution (p. 130):

$$R'COCl + HOR \rightarrow HCl(g) + R'COOR$$

Acid anhydrides react less readily and heat is usually necessary.

3. *From a silver salt and a halogenoalkane.* The two substances react when heated gently under reflux. The method is used only for the preparation of complex esters:

$$CH_3COOAg(s) + IC_2H_5(l) \rightarrow AgI(s) + CH_3COOC_2H_5(l)$$

$$R'COOAg(s) + HalR(l) \rightarrow AgHal(s) + R'COOR(l)$$

To prepare ethyl ethanoate (acetate)

Principle. Ethanoic acid is heated with excess ethanol (to ensure esterification of most of the acid). Concentrated sulphuric acid provides protons as catalyst and removes some of the water formed and thereby increases the yield of ester:

$$CH_3COOH(l) + HOC_2H_5(l) \xrightleftharpoons{H_2SO_4} H_2O(l) + CH_3COOC_2H_5(l)$$

Mixing the reagents. Mix glacial (pure) ethanoic acid (8 g) with excess ethanol (10 cm^3) in a 50 cm^3 distilling flask. To the mixture carefully add concentrated sulphuric acid (3 cm^3) and shake well during the addition. Add broken porous pot or anti-bumping granules, which prevent bumping during distillation later on.

Distilling. Arrange the reflux apparatus as in Fig. 5.2, and reflux gently for 15 minutes. Rearrange the apparatus for ordinary distillation (Fig. 6.1) and distil very slowly until about 16 cm^3 of distillate passes over. (If larger quantities are used, five times the above volumes are satisfactory in a 500 cm^3 flask, which should be heated on an oil bath at about 410 K.)

Purifying the ester. Impurities present include water, ethanol, ethanoic acid, ethoxyethane and sulphurous acid, H_2SO_3.

(a) Pour the distillate into a separating funnel. Add 10 cm^3 of 30 per cent aqueous sodium carbonate to neutralize acids. Carbon dioxide is evolved. Leave the stopper off the funnel until effervescence is no longer vigorous; then hold the stopper in firmly, with the funnel upside down, and shake well with the tap open to release the gas evolved. Finally hold the funnel upright; let the lower aqueous layer run away and retain the upper layer in the funnel.

(b) Add 10 cm^3 of 50 per cent calcium chloride solution and shake to remove ethanol. Run off the lower aqueous layer.

(c) Transfer the ester to a dry test-tube. Add anhydrous calcium chloride. Either shake for 10 minutes or allow to stand until the next lesson.

(d) Pour the dry ester into a distilling flask. Distil in the usual way. Ethoxyethane usually passes off first. Collect the ester (b.p. 350 K) which passes over between 344 and 350 K. The yield of 8 g is about 66 per cent of the theoretical yield.

Properties of esters

Physical properties. Esters of alkanoic acids are colourless oily liquids. Their odours are pleasant and resemble those of fruits, for example, the smells of ethyl ethanoate, pentyl ethanoate and methyl butanoate resemble those of apples, pear drops and pineapples respectively. Ethyl benzoate smells of peppermints. Ethyl ethanoate is only slightly soluble in water. The solubilities decrease and the melting points and boiling points increase with relative molecular mass, i.e. with the length of the carbon chain. The higher esters are oils or waxy solids.

Chemical properties

1. *Hydrolysis.* Water hydrolyses esters very slowly, even on warming, and an equilibrium is reached:

$$CH_3COOC_2H_5(l) + H_2O \rightleftharpoons CH_3COOH(aq) + C_2H_5OH(aq)$$

Acids act as sources of protons, which catalyse the reaction. Complete hydrolysis occurs when esters are warmed with aqueous or alcoholic sodium (or potassium) hydroxide because the alkali neutralizes the acid formed:

$$CH_3COOC_2H_5(l) + OH^-(aq) \rightarrow CH_3COO^-(aq) + C_2H_5OH(aq)$$

Saponification is the hydrolysis of an ester to produce an alcohol and the sodium or potassium salt of an acid:

$$RCOOR' + NaOH \rightarrow RCOONa + R'OH$$

2. *Ammonia.* Esters react with concentrated ammonia, usually in alcoholic solution, to form an amide:

$$CH_3COOC_2H_5 + NH_3 \rightarrow C_2H_5OH + \underset{\text{ethanamide}}{CH_3CONH_2}$$

3. *Reduction.* Lithium tetrahydridoaluminate(III), in dry ether as a solvent, rapidly reduces esters to alcohols:

$$CH_3COOC_2H_5 \xrightarrow{Li[AlH_4]} CH_3CH_2OH$$

This reducing agent replaces the carbonyl group of acids and acid derivatives with a CH_2 group:

$$RCOCl \text{ or } (RCO)_2O \text{ or } RCOOR' \text{ or } RCOOH \xrightarrow{Li[AlH_4]} RCH_2OH$$

Uses of esters

Esters are used in synthetic perfumes and as flavouring substances in foods. They are solvents for lacquers. Rayon (artificial silk) is cellulose ethanoate, and Terylene is an ester or, to be exact, a polyester. Waxes, fats and oils are esters, and some are used to make margarine, soap, candles, and synthetic lard. Linseed oil is the commonest drying oil in paints.

Terylene

This fibre is used for making suits and other clothing. Its texture is similar to that of wool, especially when it is mixed with about 45 per cent of wool. Unlike wool, it does not absorb water and therefore it can be used to make drip-dry clothing. Permanent creases can be made in Terylene trousers and other clothes and they last the lifetime of the cloth. Terylene is not affected by sunlight and therefore it is suitable for curtains. It is also used for making fishing lines, sails, conveyor belts and similar articles.

Terylene is made from ethane-1,2-diol, $HOCH_2CH_2OH$, and benzene-1,4-dicarboxylic acid, $HOOCC_6H_4COOH$. Many molecules join together to form a macromolecule and molecules of water are eliminated. The process is called *condensation polymerization* or *polycondensation.* A molecule of the diol has two reactive OH groups, one at each end, and the dicarboxylic acid has two reactive COOH groups. These groups react to form a polyester:

$$\begin{array}{c} n(HOCH_2CH_2OH) \\ + \\ n(HOOCC_6H_4COOH) \end{array} \rightarrow \left[O-CH_2-CH_2-O-OC-C_6H_4-CO \right]_n + nH_2O$$

If the diol is represented by HO—X—OH and the acid by HOOC—Y—COOH, the structure of Terylene is clearer:

$$-O-X-O-\underset{\underset{O}{\|}}{C}-Y-\underset{\underset{O}{\|}}{C}- \qquad \begin{array}{l} X \text{ is } CH_2CH_2 \\ Y \text{ is } C_6H_4 \end{array}$$

Fats and oils

A fat is a solid at room temperatures and an oil is a liquid. This distinction is not rigid and a substance may be an oil in hot countries and a fat in cold countries.

Animal fats include beef fat, mutton fat, butter and lard. Animal oils include fish oils such as cod liver oil. Vegetable oils, obtained by crushing the seeds or fruits of plants, include coconut oil, palm oil, linseed oil, olive oil, and oils from groundnuts, cotton seed, sunflower seed and soya beans.

Fats and oils are esters with fairly high relative molecular masses. They are formed from propane-1,2,3-triol (glycerol or glycerine), $CH_2OHCHOHCH_2OH$, and mainly two long-chain alkanoic acids, $CH_3(CH_2)_{14}COOH$ and $CH_3(CH_2)_{16}COOH$, and one unsaturated acid, $CH_3(CH_2)_7CH{=}CH(CH_2)_7COOH$. The acids are hexadecanoic acid, octadecanoic acid, and octadec-9-enoic acid. Common names of the acids are palmitic, stearic and oleic acids.

$C_{15}H_{31}COOCH_2$	$C_{17}H_{35}COOCH_2$	$C_{17}H_{33}COOCH_2$
|	|	|
$C_{15}H_{31}COOCH$	$C_{17}H_{35}COOCH$	$C_{17}H_{33}COOCH$
|	|	|
$C_{15}H_{31}COOCH_2$	$C_{17}H_{35}COOCH_2$	$C_{17}H_{33}COOCH_2$
propane-1,2,3-triyl trihexadecanoate (glyceryl tripalmitate)	propane-1,2,3-triyl trioctadecanoate (glyceryl tristearate)	propane-1,2,3-triyl trioctadecenoate (glyceryl trioleate)
$(C_{15}H_{31}COO)_3C_3H_5$	$(C_{17}H_{35}COO)_3C_3H_5$	$(C_{17}H_{33}COO)_3C_3H_5$

The above esters are simple ones, but most fats and oils consist of mixed esters such as $(C_{17}H_{35}COO)_2(C_{15}H_{31}COO)C_3H_5$. Fats contain a relatively large proportion of the saturated esters; as the proportion of unsaturated ester increases the substance becomes soft and even liquid. Most oils contain much unsaturated ester.

Manufacture of margarine

Vegetable oils contain a large proportion of the unsaturated octadecenoate (oleate) group and are unsaturated. The oils are not useful as food. They combine with hydrogen under about 5 atm pressure and with nickel as catalyst at 470 K. Hydrogen adds to the double bonds and forms saturated octadecanoate groups. The oils become solid fats, and the process is called the *hydrogenation* or *hardening of oils.* It can be stopped at any time to yield a product of suitable consistency.

$$(C_{17}H_{33}COO)_3C_3H_5(l) + 3H_2(g) \rightarrow (C_{17}H_{35}COO)_3C_3H_5(s)$$

or

$$3(-CH{=}CH-) + 3H_2 \rightarrow 3(-CH_2-CH_2-)$$

Milk, vitamins A and D, and colouring matter are added to form margarine, a substitute for butter.

A synthetic lard is made by partial hydrogenation of certain oils, and a fat useful for making candles is obtained by complete hydrogenation.

Manufacture of soap

Soap is the sodium or potassium salts of organic acids in fats and oils. It is made by hydrolysis or *saponification* of fats (beef fat, mutton fat, or fats obtained by hydrogenation of oils) with hot sodium (or potassium) hydroxide:

$$\text{Fat or oil} + NaOH(aq) \rightarrow \text{Soap}(aq) + \text{Propane-1,2,3-triol}(aq)$$

The soap is in the homogeneous solution which results, and it is precipitated by addition of concentrated sodium chloride. This process is called 'the salting out of soap'. The soap forms a curd and the triol (glycerol or glycerine) remains in the lower aqueous layer.

Superheated steam is used instead of aqueous alkali in a modern method of hydrolysing fats. The free acids are formed and they are neutralized by alkali to form soap. The 'salting out' process is not needed.

Hard soap consists of sodium salts, mainly $C_{15}H_{31}COONa$ and $C_{17}H_{35}COONa$. Soft soap consists of the potassium salts, and they also contain some $C_{17}H_{33}COOK$ and the triol. They are more soluble in water.

Liquid soap is a mixture of soft soap and coconut oil. Soap powders consist of powdered soap and sodium carbonate-10-water.

Detergents

A detergent is any substance used for washing away dirt and grease. The commonest detergent is soap and for centuries it was the only one. In everyday life the word detergent means certain synthetic products used as substitutes for soap.

How soap removes grease. The formula of one salt in ordinary soap is $C_{17}H_{35}COONa$ or $(C_{17}H_{35}COO^- + Na^+)$. The ionic COO^- group is at the end of an unbranched chain of 17 carbon atoms. The COO^- group readily mixes with water and is called *hydrophilic* (water-loving). The long hydrocarbon chain $C_{17}H_{35}$ does not mix with water and is *hydrophobic* (water-hating), but it mixes readily with grease and is fat-soluble. Soap can therefore mix with both water and grease. The hydrocarbon chains of soap penetrate grease on cloth, and the COO^- groups remain in the water. The grease splits up into an emulsion of tiny drops and leaves the cloth. Dirt is usually held to cloth by grease, and therefore removing the grease also removes the dirt.

Synthetic or soapless detergents. Most detergents are made from alkylbenzenes obtained during the refining of petroleum. They are sulphonated by concentrated sulphuric acid or sulphur trioxide, and the sodium alkylbenzenesulphonates are detergents:

$$\underset{\text{alkylbenzene}}{RC_6H_5} \xrightarrow{SO_3} \underset{\text{sulphonic acid}}{RC_6H_4SO_2OH} \xrightarrow{NaOH} \underset{\text{detergent}}{RC_6H_4SO_2O^-Na^+}$$

The alkyl group R contains between 10 and 18 carbon atoms. If it contains more than 18, the sodium salt is not soluble enough in water; if it contains less than 10, the alkyl group is not soluble enough in grease.

The formulae of two kinds of detergents can be represented as below, in which the alkyl group is a zig-zag line:

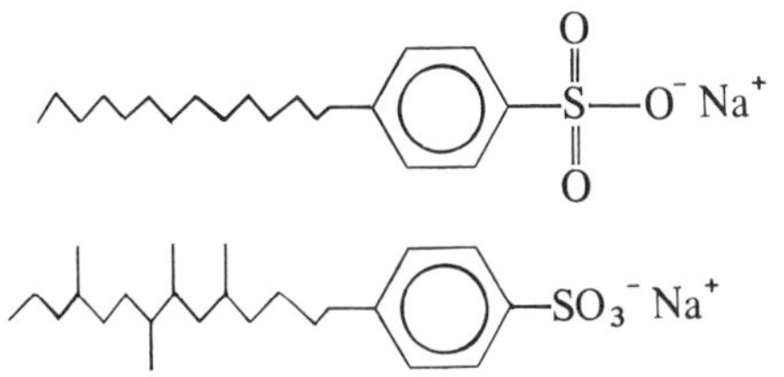

$$\text{(zig-zag)}-C_6H_4-S(=O)_2-O^-\ Na^+$$

$$\text{(branched zig-zag)}-C_6H_4-SO_3^-\ Na^+$$

Note the side-chains in the second formula. Detergent molecules with such chains cannot be broken down by bacteria and are called non-biodegradable. Foam from such detergents used to pollute some streams and rivers. Molecules with unbranched alkyl groups are biodegradable and are destroyed by bacteria.

Non-ionic detergents are made from alkylphenols, RC_6H_4OH. Epoxyethane, CH_2CH_2O, adds to the OH group, forming $RC_6H_4OCH_2CH_2OCH_2CH_2\ldots OCH_2CH_2OH$ or $RC_6H_4(OCH_2CH_2)_nOH$ or

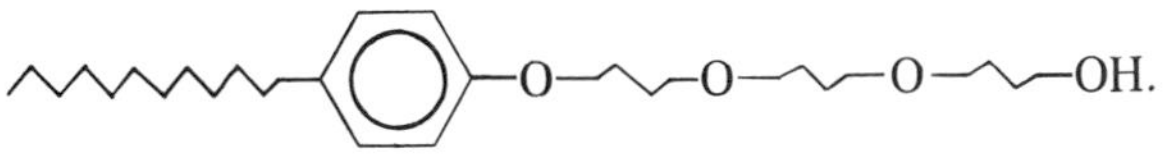

$$\text{(zig-zag)}-C_6H_4-O-\wedge-O-\wedge-O-\wedge-OH.$$

The RC_6H_4 is fat-soluble and the $(OCH_2CH_2)_nOH$ is water-soluble. These compounds are in liquid soapless detergents, and they clean even in the presence of acid, alkali and other electrolytes.

Detergents are more soluble in water than soap, and they do not form scum with hard water because their calcium and magnesium salts are soluble. Soap forms a scum because the calcium and magnesium salts of alkanoic acids are insoluble.

SUMMARY

Esters

R′COOR (R, R′ = alkyl or aryl group; R may also be H). Functional group:

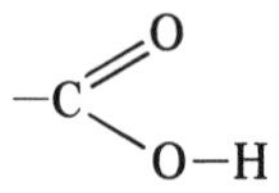

$$-C\begin{matrix}\nearrow O \\ \searrow O-H\end{matrix}$$

Preparation

1. Carboxylic acid + alcohol, $H^+(aq)$ as catalyst.

2. Acid chloride or acid anhydride + alcohol or phenol.

Properties

1. Hydrolyse to carboxylic acid and alcohol or phenol. Slow with water, rapid with OH^-(aq or alc).

2. NH_3(alc) forms amides.

3. Reduction with $Li[AlH_4]$ (ether) forms alcohols.

Terylene

Manufactured from ethane-1,2-diol, $HOCH_2CH_2OH$, and benzene-1,4-dicarboxylic acid, $HOOCC_6H_4COOH$. Simple formula of the polyester is $\left[O-X-O-CO-Y\right]_n$ ($X = CH_2CH_2$, $Y = C_6H_4$).

Fats and oils

A fat is a solid, an oil is a liquid. They are esters of propanetriol (glycerol), $CH_2OHCHOHCH_2OH$. Fats are saturated but oils are usually unsaturated and have a double bond. Vegetable oils are esters of octadecenoic (oleic) acid: $CH_3(CH_2)_7CH{=}CH(CH_2)_7COOH$.

Margarine

H_2/Ni under pressure at 200 °C saturates oil molecules (hydrogenation or hardening of oils) and forms fats. Addition of milk, vitamins and colouring matter produces margarine. Lard and synthetic candle wax are also made by hydrogenation.

Soap

Produced by alkaline hydrolysis of fats. Soaps are Na or K salts of palmitic, stearic and oleic acids.

Soapless detergents

These are Na salts of alkylbenzenesulphonic acids.

$$\text{Petroleum} \xrightarrow{\text{refined}} \underset{\text{alkylbenzene}}{RC_6H_5} \xrightarrow[\text{or } H_2SO_4]{SO_3} \underset{\text{alkylbenzene-sulphonic acid}}{RC_6H_4SO_2OH} \xrightarrow{\text{NaOH}} \underset{\text{detergent}}{RC_6H_4SO_2O^-Na^+}$$

R has 10 to 18 C atoms, preferably no side-chains so that it is biodegradable.

Alkylphenols, RC_6H_4OH and epoxyethane, CH_2CH_2O, form non-ionic detergents. They have fat-soluble RC_6H_4- and water-soluble $+\!\!OCH_2CH_2\!\!+_n$ groups.

QUESTIONS

1. Describe the preparation of ethyl ethanoate (acetate) from ethanol. What precautions would you take to ensure the maximum yield of the ester from 20 cm^3 of the ethanol?

2. Explain the chemical compositions of oils, fats and soaps. How would you prepare in the laboratory a specimen of soap from a fat?

3. In what ways do the preparations and properties of esters differ from those of salts?

4. By reference to Terylene, explain the meanings of the terms ***polyester*** and ***condensation polymerization.***

5. Name one type of reaction which is common to esters, acid chlorides and acid anhydrides, and write one appropriate equation for each of the three classes of compounds. Name one type of reaction given by acid chlorides and anhydrides but not by esters, and write suitable equations.

6. The following compounds all contain the group CH_3CO-:

CH_3COOH $\qquad$ CH_3COOCH_3 $\qquad$ CH_3COCl

In what ways do the properties of the three compounds differ, despite the structural similarity? Account for the differences.

7. Write a concise essay on detergents (other than soap). Mention their preparations, structures, advantages and disadvantages.

8. The following mixtures are obtained in the usual preparation of the first-named compound in each pair: (a) benzoic acid, hydrated manganese(IV) oxide; (b) ethyl ethanoate, aqueous ethanol; (c) ethanal, water. Describe and explain the method you would use to obtain a pure sample of the compound from each of the mixtures. (C.)

9. Three isomeric compounds, ***A, B*** and ***C***, have the molecular formula $C_8H_8O_2$. (a) When heated for a long time with sodalime ***A*** gave benzene, while both ***B*** and ***C*** gave methylbenzene. (b) ***A*** was a neutral compound, whereas ***B*** and ***C*** were monobasic acids. (c) ***B*** and ***C*** were both readily chlorinated in sunlight; ***B*** gave a compound containing three chlorine atoms per molecule which was readily hydrolysed to a dibasic acid ***D*** which in turn gave an anhydride on heating. When ***D*** was heated with sodalime it gave benzene. The chlorinated product from ***C*** contained two chlorine atoms per molecule and was a strong monobasic acid. Identify ***A, B, C*** and ***D***, write equations for the reactions, and indicate how you would prepare ***A*** from benzoic acid. (C.)

10. 0.982 g of a mixture of a monobasic alkanoic acid and its ethyl ester required 10.0 cm^3 of 0.5 M sodium hydroxide to neutralize the acid, and a further 12.0 cm^3 of the same alkali to hydrolyse the ester completely. Calculate the relative molecular mass of the acid and the mass of the ester in the mixture. Identify the acid and the ester.

11. Deduce the structures of compounds ***A, B*** and ***C*** and explain, with equations, the reactions mentioned. (i) ***A***, $C_4H_6O_4$, is neutral and on hydrolysis yields an acid, $C_2H_2O_4$, which forms oxides of carbon with concentrated sulphuric acid. (ii) ***B***, $C_4H_8O_2$, is neutral. It reacts with ammonia to form compounds of molecular formulae C_2H_6O and C_2H_5NO. (iii) ***C***, $C_2H_4Br_2$, hydrolyses to a compound C_2H_4O which forms a yellow precipitate with acidified 2,4-dinitrophenylhydrazine.

12. An ester, $C_6H_{12}O_2$, was refluxed with aqueous sodium hydroxide until reaction was complete. The product was partly distilled. The distillate gave a positive haloform reaction (yellow precipitate with iodine and alkali). The residue in the distillation flask was neutralized with dilute nitric acid; with aqueous silver nitrate a silver salt containing 59.7 per cent by mass was precipitated. Deduce the structural formula of the ester.

13. The structures of esters, ethers and anhydrides are similar in that each consists of two alkyl and/or acyl groups attached to an oxygen atom. Compare and contrast in tabular form the methods of preparation, physical properties and chemical reactions of an ether, an ester and an acid anhydride. (L.)

14. For each of the following pairs of compounds suggest a chemical test which could be used to distinguish between the members of each pair. (a) CH_3CH_2I and CH_3CH_2Br. (b) $CH_3CH_2CH_2OH$ and $CH_3CH(OH)CH_3$. (c) $CH_3COOC_6H_5$ and $C_6H_5COOCH_3$. (d) $CH_3CH_2CH{=}CH_2$ and $CH_3CH_2C{\equiv}CH$. In each case, state the procedure clearly together with the expected observations and give balanced equations where appropriate. (L.)

Chapter 17
NITRILES

Nitriles contain the functional group $-C\equiv N$ and their general formula is RCN or $R-C\equiv N$ in which R is an alkyl or aryl group. They may be regarded as esters of hydrogen cyanide, HCN, and formerly they were called cyanides.

CH_3CN	CH_3CH_2CN	C_6H_5CN
ethanenitrile	propanenitrile	benzonitrile

Preparation of nitriles

1. *Dehydration of amides.* Mix an amide well with phosphorus(V) oxide, phosphorus trichloride oxide or sulphur dichloride oxide, and distil gently:

$$\underset{\text{ethanamide}}{CH_3CONH_2} \xrightarrow[PCl_3O]{P_2O_5 \text{ or}} \underset{\text{ethanenitrile}}{CH_3CN}$$

$$CH_3CONH_2(s) + SCl_2O(l) \rightarrow 2HCl(g) + SO_2(g) + CH_3CN(l)$$

$$R-\underset{\underset{O}{\|}}{C}-NH_2 \rightarrow R-C\equiv N + H_2O$$

2. *From a halogenoalkane or (halogenoalkyl) benzene.* Boil the halogen compound under reflux for about 24 hours with sodium (or potassium) cyanide in aqueous alcoholic solution; the water is a solvent for the cyanide and the alcohol for the halogen compound:

$$CH_3I(alc) + KCN(aq) \rightarrow KI + CH_3CN$$

$$C_6H_5CH_2I(alc) + KCN(aq) \rightarrow KI + \underset{\text{phenylethanenitrile}}{C_6H_5CH_2CN}$$

C_6H_5Hal does not react to form C_6H_5CN.

Reactions of nitriles

Hydrolysis.
(a) Add about 1 cm^3 of ethanenitrile to about 5 cm^3 of concentrated aqueous potassium (or sodium) hydroxide. Warm gently, preferably under reflux, and identify the gas evolved. When no more gas is evolved, add hydrochloric acid carefully until the mixture is neutral and do the iron(III) chloride test (p. 179).

(b) Add ethanenitrile to about 5 times its volume of dilute sulphuric acid, and heat gently under reflux for 30 minutes. Identify one product by its smell.

Properties of nitriles

Physical properties. Ethanenitrile is a colourless liquid (b.p. 355 K) with a pleasant sweetish odour. It is readily soluble in water but higher members are only slightly soluble.

Benzonitrile is a colourless oily liquid (b.p. 464 K) and smells of almonds.

Chemical properties. Nitriles are unsaturated and their reactions involve addition.

1. *Hydrolysis.* Mineral acids or aqueous alkalis hydrolyse nitriles when boiled together under reflux. The elements of water are probably first added across the triple bond C≡N (this reaction is a hydration), and subsequent hydrolysis of the amide produces a carboxylic acid.

$$RC{\equiv}N \rightarrow RCONH_2 \rightarrow RCOONH_4 \rightarrow RCOOH + NH_3$$

$$CH_3CN + H_2O + NaOH \rightarrow NH_3 + CH_3COONa$$

$$CH_3CN + H_2O + H_3O^+ \rightarrow NH_4^+ + CH_3COOH$$

Acid hydrolysis produces the free carboxylic acid; alkaline hydrolysis produces the sodium or potassium salt and ammonia is evolved.

$$C_6H_5CN + 2H_2O + HCl \rightarrow NH_4Cl + \underset{\text{benzoic acid}}{C_6H_5COOH}$$

2. *Reduction.* Reducing agents are (a) sodium and ethanol, (b) hydrogen in the presence of a nickel catalyst, and (c) lithium tetrahydridoaluminate(III):

$$RCN \xrightarrow{2H_2} \underset{\text{an amine}}{RCH_2NH_2}$$

The products formed by hydrolysis and by reduction show that the alkyl or aryl group is attached to the carbon atom and not to the nitrogen atom.

Lengthening a carbon chain

An additional CH_2 group can be added to a molecule by making use of the nitrile. In this way a compound can be converted to its next higher homologue. The series of reactions for converting ROH to RCH_2OH is:

$$ROH \xrightarrow{P + I_2} RI \xrightarrow{KCN} RCN \xrightarrow{reduce} RCH_2NH_2 \xrightarrow{HNO_2} RCH_2OH$$

The series for converting RCOOH to RCH_2COOH is

$$RCOOH \xrightarrow{Li[AlH_4]} RCH_2OH \xrightarrow[above]{as} RCH_2CN \xrightarrow{H_3O^+} RCH_2COOH$$

SUMMARY

Nitriles

R–CN (R = alkyl or aryl). Functional group: –C≡N.

Preparation

1. Dehydrate amides with P_2O_5 or SCl_2O. $RCONH_2 \rightarrow RCN$.
2. Reflux halogenoalkane or (halogenoalkyl)benzene with KCN(alc).

Properties

1. Hydrolysis. Refluxing with alkali or acid forms an amide.
2. Reduction with $Li[AlH_4]$ or Na + ethanol or H_2/Ni forms a primary amine.

QUESTIONS

1. Outline two methods of preparing ethanenitrile. Describe its chief physical properties. What changes occur when ethanenitrile is (a) hydrolysed and (b) reduced, and how are these reactions brought about?

2. You wish to convert ethanoic (acetic) acid to propanoic acid and to convert methanol to ethanol. Indicate how you would propose to do the conversions, mentioning the reagents and conditions necessary.

3. By means of named compounds, balanced equations, and an outline of essential conditions, describe how you would change (a) a nitrile to a ketone, (b) an alkene to an ether, and (c) an acid chloride to a hydrocarbon. (C.)

4. How would you bring about five of the following conversions in aliphatic compounds? Illustrate, giving conditions and equations with a specific named example in each case.

(a) $-NH_2 \rightarrow -OH$ (b) $-CN \rightarrow -COONa$
(c) $-CH_2 \rightarrow -COOH$ (d) $-COCl \rightarrow -CHO$
(e) $-CN \rightarrow -CH_2NH_2$

5. Give, with some practical detail, the different methods by which a hydrogen atom in the benzene molecule can be replaced by each of the following groups: $-CH_3$, $-CHO$, $-CN$, $-OC_2H_5$. (C.)

Chapter 18
NITRO-COMPOUNDS

Nitrobenzene, $C_6H_5NO_2$, is formed when benzene is nitrated at 323 K (p. 227), and at temperatures over 323 K 1,3-dinitrobenzene is the chief product.

Methylbenzene is nitrated more readily than benzene. It forms a mixture of 1-methyl-2- and 1-methyl-4-nitrobenzene, $CH_3C_6H_4NO_2$, when treated with cold dilute nitric acid. Nitration with a mixture of concentrated nitric and sulphuric acids produces 1-methyl-2,4-dinitrobenzene at about 323 K and 2-methyl-1,3,5-trinitrobenzene (trinitrotoluene or TNT) at 373 K.

Electron delocalization occurs in the nitro-group ($-NO_2$):

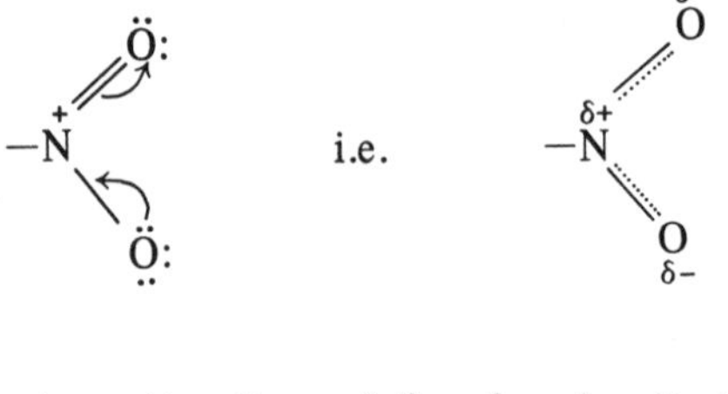

$$-\overset{+}{N}(=\ddot{O}:)(-\ddot{O}:^{-}) \quad \text{i.e.} \quad -\overset{\delta+}{N}(\overset{\delta-}{O})(\overset{\delta-}{O})$$

The nitro-group is electron-attracting and therefore deactivates a benzene ring to electrophilic attack. It is 3-(*meta*-) directing.

To prepare nitrobenzene

(Nitrobenzene is poisonous. Do not breathe the vapour or allow the liquid to touch your skin.)

Principle. A mixture of concentrated nitric and sulphuric acids contains three ions:

$$HNO_3 \text{ or } NO_2OH + HHSO_4 \rightarrow H_2O + HSO_4^- + NO_2^+$$

$$H_2O + H_2SO_4 \rightarrow H_3O^+ + HSO_4^-$$

Adding up the two equations gives:

$$HNO_3 + 2H_2SO_4 \rightarrow NO_2^+ + 2HSO_4^- + H_3O^+$$

The equation shows that addition of one molecule of nitric acid to sulphuric acid produces four particles. Evidence that this is so comes from measurements of the depression of the freezing point of sulphuric acid. The depression produced by dissolving concentrated nitric acid in sulphuric acid is four times greater than expected.

NO_2^+ the nitryl cation is the nitrating agent (p. 80):

$$C_6H_6 + NO_2^+ \rightarrow C_6H_5NO_2 + H^+$$

Mixing the reagents. Add concentrated nitric acid (10.5 cm^3) to a 50 cm^3 flask, and then add concentrated sulphuric acid (12 cm^3) a little at a time. Cool the flask during the mixing by immersing it in cold water. The acid mixture is very corrosive.

Add benzene (9 cm^3) about 1 cm^3 at a time. Throughout the addition shake the mixture vigorously to form an emulsion because benzene and the acids do not mix. Do not allow the temperature to rise above 323 K otherwise some dinitrobenzene is formed; the reaction can become uncontrollable at 333 K or above. (The masses of reagents are in the proportions: benzene 1, nitric acid 2, sulphuric acid 3.)

Refluxing. Attach a reflux condenser to the flask and heat the mixture on a water bath at about 333 K for 30 minutes. Shake the mixture well from time to time to ensure steady reaction. Allow the product to cool.

Purifying the nitrobenzene. The impurities present are the two acids used, benzene, water and possibly some dinitrobenzene.

(a) Pour the mixture, with stirring, into about 100 cm^3 of water in a beaker. Decant as much of the upper layer of aqueous acids as possible. Add more water and decant the aqueous layer once again. This treatment removes most of the acids.

(b) Transfer the nitrobenzene to a separating funnel. Add aqueous sodium carbonate (about 20 cm^3). Shake gently and release the carbon dioxide formed as described on p. 92. Decant the upper aqueous layer. Repeat the treatment with more sodium carbonate until effervescence ceases. All of the acid is removed.

(c) Run the moist nitrobenzene into a small conical flask. Its colour is cloudy. Add anhydrous sodium sulphate and shake until the liquid is a clear yellow oil.

(d) Pour the dry liquid into a distilling flask. Distil, using an air condenser. Unchanged benzene passes over at about 353 K. Collect the fraction boiling between 480 and 484 K. Some dinitrobenzene may remain in the flask.

Properties of nitro-compounds

Physical properties. Nitrobenzene is a pale yellow oily liquid (m.p. 278.8 K; b.p. 484 K) with an odour of bitter almonds. It is insoluble in water.

Chemical properties. Nitrobenzene burns readily with a smoky flame. It is not explosive. It does not react with most acids, alkalis and oxidizing agents. It can be nitrated to the 1,3-dinitro compound, and chlorine converts it to 3-nitrochlorobenzene.

Acidic reducing agents reduce nitrobenzene to phenylamine. The reagents are iron or tin and hydrochloric acid.

$$C_6H_5NO_2 + 6[H] \rightarrow 2H_2O + C_6H_5NH_2$$

$$C_6H_5NO_2 + 3Fe + 6H^+(aq) \rightarrow C_6H_5NH_2 + 3Fe^{2+}(aq) + 2H_2O$$

SUMMARY

Nitro-compounds
RNO_2 (R = aryl group). Functional group: $-NO_2$.

Preparation
$C_6H_6 + HNO_3/H_2SO_4$(conc), 50 °C → $C_6H_5NO_2$.
$C_6H_5CH_3 + HNO_3$ cold, dilute → $C_6H_2(NO_2)_3CH_3$
2-methyl-1,3,5-trinitrobenzene, TNT

Properties of nitrobenzene
Fuming HNO_3 or HNO_3/H_2SO_4(conc) over 50 °C → 1,3-dinitrobenzene.
HCl(conc) + Fe or Sn reduce to phenylamine (aniline), $C_6H_5NH_2$.

QUESTIONS

1. Describe, with full practical details, the preparation from benzene of nitrobenzene. How would you convert nitrobenzene to 1,3-dinitrobenzene?

2. What evidence is there that a mixture of concentrated nitric and sulphuric acids contains the nitryl cation NO_2^+ and other ions? Discuss the mechanism by which benzene is converted to nitrobenzene, and explain what is meant by the terms ***nucleophilic*** and ***electrophilic.***

3. (a) Benzene reacts with a mixture of concentrated nitric and sulphuric acids to form nitrobenzene. What is the nature of the species which reacts with benzene in this reaction and how is it formed? (b) Reagents which form addition products with propene do not usually form addition products with ethanal and vice versa. Explain why this is so.

4. Three isomers of the formula $NO_2C_6H_4COOH$ exist. Write down their full displayed formulae and names. State which of these nitro-compounds you would expect to be formed by nitration of benzoic acid, C_6H_5COOH. Outline how you would prepare specimens of the other isomer(s).

5. Monochlorination of benzene in the presence of iron(III) chloride followed by nitration of the product gives two isomers of formula $ClC_6H_4NO_2$. Write displayed (graphic) formulae for these isomers, and suggest a method by which a third isomer could be prepared from benzene. By what reaction would you distinguish between isomers of formulae: $NH_2C_6H_4OCH_2COOC_2H_5$, $OHC_6H_4NHCH_2COOC_2H_5$ and $OHC_6H_4CH(NH_2)COOC_2H_5$? (C.)

Chapter 19
AMINES

Amines may be regarded as derivatives of ammonia, in which 1, 2 or 3 of the hydrogen atoms have been replaced by alkyl or aryl groups. The functional groups present are $-NH_2$ in primary amines, $=NH$ in secondary amines, and $\equiv N$ in tertiary amines. Examples are:

primary amine	secondary amine	tertiary amine
$R-NH_2$	R_2NH or $R_2>NH$	R_3N or $R, R', R'' \equiv N$
CH_3NH_2 or CH_3-NH_2 (N bonded to H, H)	$(CH_3)_2NH$ or $(CH_3)(CH_3)>N-H$	$(CH_3)_3N$ or $(CH_3)(CH_3)(CH_3)\equiv N$
methylamine	dimethylamine	trimethylamine

The above are simple amines. Secondary and tertiary mixed amines also exist:

$$\begin{matrix} CH_3 \\ C_2H_5 \end{matrix} > N-H \qquad\qquad \begin{matrix} CH_3 \\ CH_3 \\ C_2H_5 \end{matrix} \equiv N$$

methylethylamine ethyldimethylamine

Some primary amines are:

Name	*Formula*	*B.p./K*
Methylamine	CH_3NH_2	266.8
Ethylamine	$CH_3CH_2NH_2$	289.7
Propylamine	$CH_3CH_2CH_2NH_2$	321.7
Butylamine	$CH_3CH_2CH_2CH_2NH_2$	350.6
Phenylamine (aniline)	$C_6H_5NH_2$	457.3

Isomerism of amines. There are four isomers C_3H_9N:

$CH_3CH_2CH_2NH_2$ — propylamine

$CH_3CH(NH_2)CH_3$ — (1-methylethyl)amine

$(CH_3)(C_2H_5)NH$ — ethylmethylamine

$(CH_3)_3N$ — trimethylamine

Preparation of primary amines

1. *From ammonia and halogenoalkane (Hofmann's reaction).* Heat the halogen compound (usually the iodine derivative because it is most reactive and least volatile) at about 373 K under pressure in a sealed tube with excess aqueous or alcoholic ammonia:

$$C_2H_5I(alc) + NH_3(alc) \rightarrow C_2H_5NH_3^+I^-(s)$$

Hot alkali liberates the free amines from the products.

Phenylamine cannot be prepared in the laboratory by this method because the halogen in C_6H_5Hal is held too firmly. However, it is manufactured commercially by carrying out the reaction under extreme conditions (470 K, 60 atm and a catalyst).

2. *Reduction of a nitrile or amide.* Reduce a nitrile or amide with lithium tetrahydridoaluminate(III)

$$RCN \rightarrow RCH_2NH_2; \qquad RCONH_2 \rightarrow RCH_2NH_2$$

$$C_6H_5CN + 4[H] \rightarrow C_6H_5CH_2NH_2$$

3. *By action of bromine and potassium hydroxide on an amide.* Mix the amide with bromine. Add cold dilute potassium hydroxide and shake until the reddish brown colour of bromine disappears and the mixture is yellowish. Add concentrated potassium hydroxide and warm at about 340 K:

$$\underset{\text{ethanamide}}{CH_3CONH_2} + Br_2 + 4KOH \rightarrow 2KBr + K_2CO_3 + CH_3NH_2$$

The reaction takes place in stages and involves an intramolecular rearrangement of C—C—N to C—N—C with dilute alkali:

$$\underset{\text{amide}}{RCONH_2} \xrightarrow{Br_2} \underset{N\text{-bromoamide}}{RCONHBr} \xrightarrow[KOH]{\text{dil}} \underset{\text{isocyanate}}{RNCO} \xrightarrow[KOH]{\text{conc}} \underset{\text{amine}}{RNH_2}$$

The change of the amide to an amine results in the loss of one carbon atom per molecule; the reaction is used to pass from a homologue to the next lower compound in the series.

4. *Reduction.* Nitroarenes are reduced by acid reducing agents:

$$C_6H_5NO_2 \xrightarrow[\text{Sn or Fe}]{HCl} \underset{\text{phenylamine}}{C_6H_5NH_2}$$

To prepare phenylamine (aniline)

Principle. Nitrobenzene is reduced with hydrogen from hydrochloric acid and either iron or tin:

$$C_6H_5NO_2 \xrightarrow{+6[H]} C_6H_5NH_2 + 2H_2O$$

Mixing the reagents for reduction. To a 500 cm^3 flask add iron filings (20 g), water (40 cm^3) and concentrated hydrochloric acid (40 cm^3). Fit a reflux air condenser to the flask. Add nitrobenzene (10 cm^3) in small portions at a time down the condenser tube. Either stir mechanically or shake vigorously between each addition to ensure mixing of the reagents. When all the nitrobenzene has been added, heat the flask on a water bath until the almond-like smell of nitrobenzene has disappeared. Phenylammonium chloride is in solution.

Liberation of phenylamine. Allow the apparatus to cool. Remove the condenser. Carefully add a paste of calcium hydroxide or concentrated aqueous sodium hydroxide to the flask until the mixture is strongly alkaline. The amine separates as a brown oil:

$$C_6H_5NH_3^+(aq) + OH^-(aq) \rightarrow H_2O + C_6H_5NH_2(l)$$

Steam distillation. Phenylamine solution 'bumps' very much at its boiling point and therefore cannot be distilled in the ordinary way. Steam distillation, in the apparatus of Fig. 19.1, is used; the amine passes over at about 372 K and no bumping occurs.

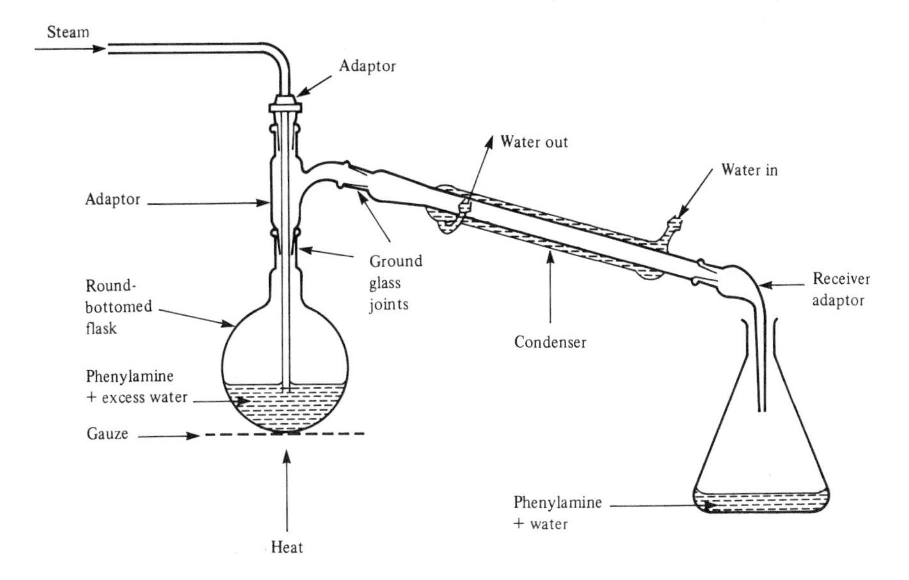

Fig. 19.1. Steam distillation of phenylamine

Pass steam into the mixture, heated on a gauze or sand tray. the milky distillate consists of the amine and water. Continue the distillation until only water passes over as a clear distillate.

Extraction with ether. Add solid sodium chloride to the distillate until it is saturated, because the amine is much less soluble in sodium chloride solution than in water. Add the mixture of liquids to a separating funnel. Shake with about 20 cm^3 of ether (all flames must be extinguished); lift the stopper from time to time to release the pressure. Run off the lower layer into a beaker and the upper layer (amine + ether) into a flask. Return the liquid from the beaker to the funnel and again extract with ether; add the second ether extract to the first one. Repeat the extraction process with a third portion of ether.

Drying the amine. To the ether extracts add either sodium (or potassium) hydroxide pellets or anhydrous sodium sulphate. Cork the flask and let it stand for one day. (Phenylamine reacts with sulphuric acid, calcium chloride and phosphorus(V) oxide, and they cannot be used to dry it.)

Distillation. Distil off the ether on a water bath. Distil the residue, using an air condenser, and collect the fraction passing over between 455 and 457 K.

The residue, which is almost pure phenylamine, does not 'bump' much.

Reactions of primary amines

1. *Odour.* Warm a little methylammonium chloride with dilute sodium hydroxide. Smell the methylamine gas evolved.

2. *Solubility and basic properties.* Shake butylamine, phenylamine and (phenylmethyl)amine with water in separate test-tubes. Note their relative solubilities. Add 2 drops of universal indicator to the solutions; estimate their relative basic strengths.

3. *Combustion.* Add a few drops of the amines to separate watch glasses and find whether they burn in air.

4. *Nitrous acid, HNO_2.*
(a) Add dilute hydrochloric acid to aqueous ethylammonium chloride, and then add a few crystals of sodium nitrite, $NaNO_2$. Note the gas evolved.

(b) Dissolve about 0.5 cm^3 of phenylamine in 10 cm^3 of dilute hydrochloric acid. Add the solution to about 25 cm^3 of crushed ice and water at about 278 K. Now add, in small portions at a time, a solution of 0.5 g of sodium nitrite in 10 cm^3 of water, and stir constantly during the addition. Divide the product into two parts. Warm one part; use the other part in the next test (within 5 minutes).

5. *Naphthalen-2-ol.* Dissolve 1 g of naphthalen-2-ol in 15 cm^3 of dilute sodium hydroxide, warming if necessary. Add cold water until the volume is 50 cm^3. Now add small portions of the reaction mixture from test 4(b). Note the colour of the product.

6. *Salts of metals.* Add aqueous copper(II) sulphate and iron(III) chloride separately, drop by drop, to an aqueous solution of an amine. Observe the changes.

7. *Bromine and phenylamine.* Dissolve 2 drops of phenylamine in dilute sulphuric acid. Add bromine water and observe any changes.

8. *Acid chlorides and anhydrides.* Refer to p. 195.

Properties of primary aliphatic amines

Physical properties. Methylamine is a gas at room temperature and ethylamine is a volatile liquid; (phenylmethyl)amine is a colourless liquid. Aliphatic amines have pungent, fishy smells similar to that of ammonia. The lower members are readily soluble in water.

Chemical properties

1. *Combustion.* They burn in air with a yellow flame, whereas ammonia does not burn in air but burns in oxygen. This is the simplest property that distinguishes methylamine from ammonia.

2. *Basic properties.* 1 cm^3 of water dissolves more than 1000 cm^3 of methylamine at room temperatures. Aqueous solutions of amines are alkaline:

$$CH_3NH_2(g) + H_2O \rightleftharpoons CH_3NH_3^+(aq) + OH^-(aq)$$

The carbon-nitrogen bond is polar: $\overset{\delta+}{C}-\overset{\delta-}{N}$. An alkyl group is electron-releasing (relative to a hydrogen atom). The substitution in ammonia of alkyl for hydrogen makes the lone pair more readily available, and alkylamines are more basic than ammonia. The phenyl group, C_6H_5, is electron-attracting, and therefore phenylamine is less basic than ammonia and other amines (p. 236).

Amines and mineral acids react readily and form crystalline salts similar to ammonium salts:

$$CH_3NH_3^+Cl^- \qquad C_2H_5NH_3^+NO_3^- \qquad (CH_3NH_3^+)_2SO_4^{2-}$$

Alkalis liberate the free amines from the salts.

Amines resemble ammonia solution in precipitating hydroxides from aqueous solutions of metallic salts and in forming complex ions with copper:

$$Fe^{3+}(aq) + 3OH^-(aq) \rightarrow Fe(OH)_3(s), \text{ reddish brown}$$

2. *Acid chlorides and acid anhydrides.* Amines are ethanoylated readily at room temperatures by ethanoyl chloride and less readily by ethanoic anhydride:

$$CH_3COCl + H_2NCH_3 \rightarrow HCl + \underset{N\text{-methylethanamide}}{CH_3CONHCH_3},$$

Benzoyl chloride, C_6H_5COCl, reacts similarly (p. 197).

3. *Nitrous acid, HNO_2.* The acid itself is so unstable that it cannot be used. It is produced from its sodium salt and a mineral acid. Primary aliphatic amines form nitrogen gas and compounds R—X where X = OH, NO_2 or ONO, and the compounds are an alcohol, nitro-compound or nitrite respectively:

$$CH_3NH_2 + ONOH \rightarrow H_2O + N_2(g) + \underset{\text{methanol}}{CH_3OH}$$

$$CH_3NH_2 + HNO_2 \rightarrow H_2O + N_2(g) + \underset{\text{methyl nitrite}}{CH_3NO_2}$$

$$CH_3NH_2 + HNO_2 \rightarrow H_2O + N_2(g) + \underset{\text{nitromethane}}{CH_3NO_2}$$

A diazonium cation, $CH_3{-}N{\equiv}N^+$, is first formed, and it at once decomposes into nitrogen and a carbocation, CH_3^+, which then combines with an electron pair donor (OH, NO_2 or ONO) to form the final products.

Properties of phenylamine (aniline), $C_6H_5NH_2$

Physical properties. Phenylamine is a colourless oily liquid (m.p. 267 K, b.p. 457 K). It has a characteristic odour and is poisonous. It is only slightly soluble in water, forming a 3.5 per cent solution at 298 K, but dissolves in most organic solvents. It turns dark on exposure to air and light.

The nitrogen atom in the amino group $-NH_2$ has an unshared electron pair and makes the group a nucleophile:

```
      H
      ..
    R:N:  ← unshared electron pair
      ..
      H
```

Therefore primary amines (and other amines) act as bases (proton acceptors).

The aromatic amine phenylamine, $C_6H_5NH_2$, is a much weaker base than aliphatic amines because the unshared electron pair of the nitrogen is delocalized with the π-electrons of the benzene ring:

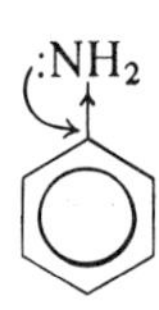

As a result the electron pair is less readily available to a proton. Although the nitrogen atom is electronegative the delocalization effect is stronger. The net effect is that the ring is more reactive towards electrophiles (contrast nitrobenzene and phenol). The 2- and 4- (*ortho* and *para*) positions are activated.

Chemical properties. Phenylamine resembles aliphatic amines in its reactions with water, acids, acid chlorides, acid anhydrides, warm nitrous acid, and trichloromethane.

1. *Basic properties.* It is a weaker base than ammonia and aliphatic amines, and its aqueous solution does not turn litmus fully blue:

$$C_6H_5NH_2(l) + H_2O \rightleftharpoons C_6H_5NH_3^+(aq) + OH^-(aq)$$

It forms crystalline salts with acids. With hydrochloric acid it forms a solid mass of the chloride, which is soluble in water; the sulphate is sparingly soluble:

$$C_6H_5NH_3^+Cl^- \qquad (C_6H_5NH_3^+)_2SO_4^{2-}$$

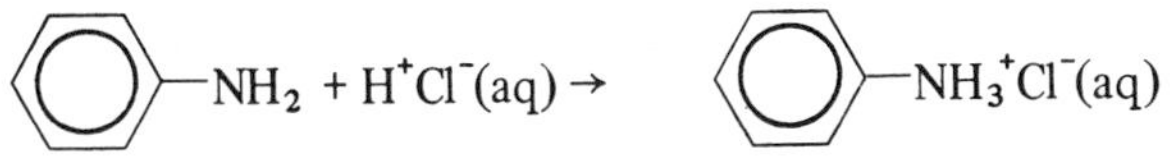

$$C_6H_5-NH_2 + H^+Cl^-(aq) \rightarrow C_6H_5-NH_3^+Cl^-(aq)$$

Aqueous phenylamine precipitates hydroxides of iron, aluminium and zinc from their salts.

2. *Acid chlorides and anhydrides.* CH_3CO- and C_6H_5CO- derivatives (p. 198) are formed when phenylamine reacts with ethanoyl chloride at room temperature, and with warm ethanoic anhydride, and benzoyl chloride, C_6H_5COCl:

$CH_3CONHC_6H_5$ — *N*-phenylethanamide

$C_6H_5CONHC_6H_5$ — *N*-phenylbenzamide

3. *Alkylation and arylation.* Halogenoalkanes replace a hydrogen atom by an alkyl group:

$$C_6H_5NH_2 + C_2H_5Br \rightarrow C_6H_5NH_2C_2H_5^+\,Br^-$$

ethylphenylammonium bromide

4. *Nitrous acid, HNO_2.* A benzenediazonium salt is produced at temperatures between 273 and 278 K. These are more stable than the very unstable aliphatic diazonium salts.

$$C_6H_5NH_2 \xrightarrow[\text{(acid)}]{HNO_2} C_6H_5-N\equiv N^+$$

benzenediazonium ion

Nitrogen and a hydroxy-compound are formed by hydrolysis when the diazonium salt is warmed (this change is similar to that with aliphatic amines).

$$C_6H_5N_2Cl(aq) + H_2O \rightarrow N_2(g) + C_6H_5OH(aq) + HCl(aq)$$

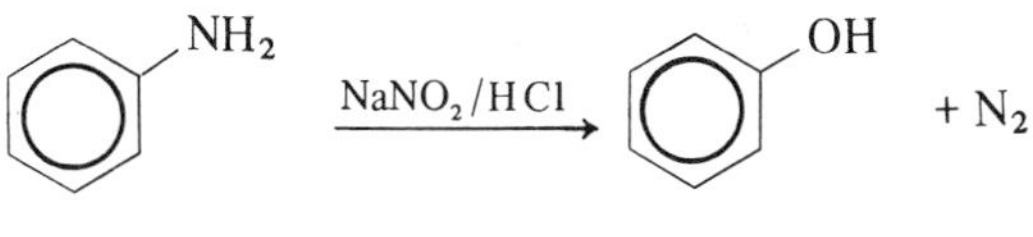

$$C_6H_5NH_2 \xrightarrow{NaNO_2/HCl} C_6H_5OH + N_2$$

5. *Reactions involving the benzene ring.* The NH_2 group, like the OH group, increases the chemical reactivity of the ring.

(a) *Halogenation.* Bromine water and chlorine water at once form white precipitates:

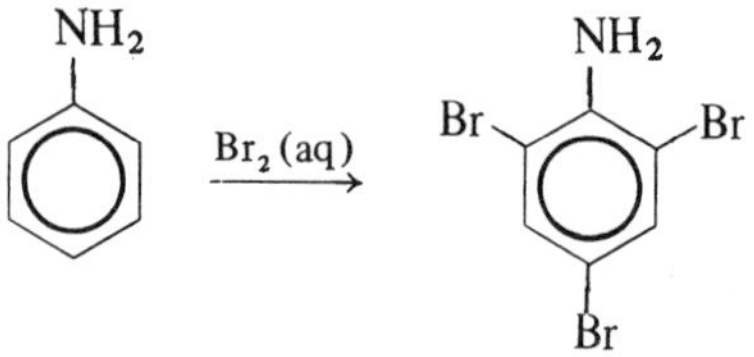

$$C_6H_5NH_2 \xrightarrow{Br_2(aq)} 2,4,6\text{-}Br_3C_6H_2NH_2$$

2,4,6-tribromophenylamine

(b) *Nitration.* Nitric acid oxidizes and chars the amine and forms no useful product.

(c) *Sulphonation.* Excess concentrated sulphuric acid forms phenylammonium hydrogensulphate; on heating to 450 K this changes to 4-aminobenzenesulphonic acid:

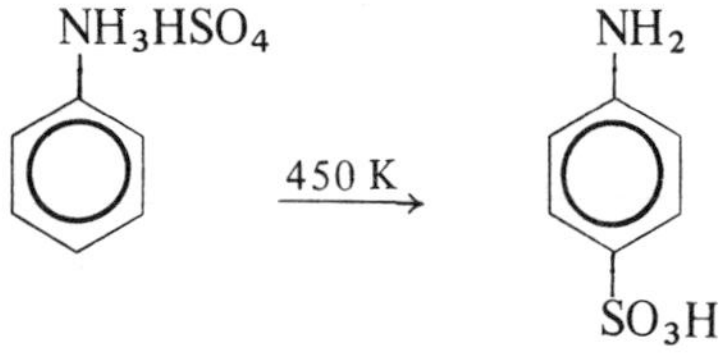

$$C_6H_5NH_3HSO_4 \xrightarrow{450\ K} 4\text{-}H_2NC_6H_4SO_3H$$

Diazotization

Aqueous sodium nitrite, $NaNO_2$, reacts at 278 K with phenylamine, acidified with excess acid:

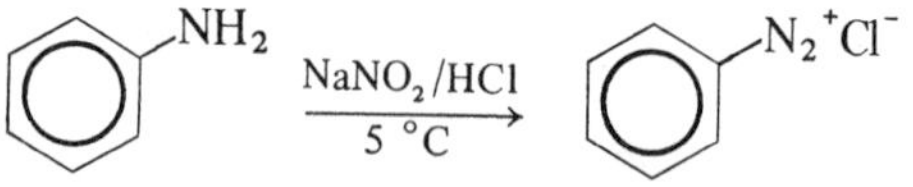

$$C_6H_5NH_2 \xrightarrow[5\ ^\circ C]{NaNO_2/HCl} C_6H_5N_2^+Cl^-$$

a diazonium salt

The reaction is called diazotization. The sodium nitrite is added until starch-iodide paper is turned blue (after allowing suitable time for the diazotization to occur). The blue colour indicates that excess nitrous acid is present. Diazotization is slow below 278 K, and at 283 K and above phenol and nitrogen are formed. Nitrogen is evolved rapidly at 340 K:

$$C_6H_5N_2Cl(aq) + H_2O \rightarrow C_6H_5OH(l) + N_2(g) + HCl(aq)$$

(Phenylmethyl)amine, $C_6H_5CH_2NH_2$, does not diazotize because its NH_2 group is not attached directly to a benzene nucleus.

Diazonium salts are used in solution because the solids are unstable and explosive. The $-N_2$ group can be replaced by a halogen, and benzenediazonium compounds form (a) chlorobenzene with warm hydrochloric acid and copper(I) chloride, (b) bromobenzene with warm hydrobromic acid and copper(I) bromide, and (c) iodobenzene with warm aqueous potassium iodide:

$$C_6H_5N_2^+Cl^- \xrightarrow[HCl]{CuCl} N_2(g) + \underset{\text{chlorobenzene}}{C_6H_5Cl}$$

$$C_6H_5N_2^+Cl^- \xrightarrow[HBr]{CuBr} N_2(g) + \underset{\text{bromobenzene}}{C_6H_5Br}$$

$$C_6H_5N_2^+Cl^- \xrightarrow{KI(aq)} N_2(g) + \underset{\text{iodobenzene}}{C_6H_5I}$$

These compounds cannot be prepared directly from benzene and the halogen. Note that chlorobenzene and bromobenzene cannot be prepared by using aqueous potassium chloride or bromide.

Coupling. In the presence of alkali, diazonium salts join or couple with phenols and aromatic amines to form coloured compounds, containing the group C$-$N$=$N$-$C, often used as dyes. Phenol forms bright orange 4-(phenylazo)phenol (the $-N{=}N-$ group is the azo group):

$$C_6H_5{-}N_2^+Cl^- + C_6H_5{-}OH \xrightarrow{NaOH} C_6H_5{-}N{=}N{-}C_6H_4{-}OH$$

Naphthalen-2-ol at once forms a bright red precipitate, and this reaction is a good test for phenylamine.

naphthalen-2-ol ($C_{10}H_7OH$)

SUMMARY

Amines

General formula: $R{-}NH_2$ (R = alkyl or aryl group). Functional group of primary amines: $-NH_2$.

Preparation

1. Halogenoalkane with NH_3(aq or alc) under pressure.
2. Reduce a nitrile or amide with Li[AlH_4] (ether).
3. Br_2(l) and KOH(aq) on an amide.
4. Reduce nitrobenzene with Fe/HCl or Sn/HCl → $C_6H_5NH_2$ phenylamine (aniline)

Properties

Reagent/reaction/ property	*Ethylamine, $C_2H_5NH_2$*	*Phenylamine, $C_6H_5NH_2$*
Physical properties	Volatile liquid Lower b.p. Soluble in water	Oily liquid Higher b.p. Sparingly soluble
Basic properties Acids form salts	More alkaline than NH_3(aq) $C_2H_5NH_3^+Cl^-$(s), white	Less alkaline than NH_3(aq) $C_6H_5NH_3^+Cl^-$(s), white
Alkylation, CH_3I	$C_2H_5NHCH_3$ secondary amine	$C_6H_5NHCH_3$ secondary amine
Ethanoylation, CH_3COCl	$CH_3CONHC_2H_5$ *N*-ethylethanamide	$CH_3CONHC_6H_5$ *N*-phenylethanamide
Benzoylation, C_6H_5COCl	$C_6H_5CONHC_2H_5$ *N*-ethylbenzamide	$C_6H_5CONHC_6H_5$ *N*-phenylbenzamide
HNO_2, warm	$C_2H_5OH + N_2$ ethanol	$C_6H_5OH + N_2$ phenol
HNO_2, below 10 °C	$C_2H_5OH + N_2$	$C_6H_5{-}N{=}N^+$ diazonium ion
Halogenation, Br_2(aq), Cl_2(aq)	No reaction	e.g. $C_6H_2Br_3NH_2$(s), white 2,4,6-tribromophenylamine
H_2SO_4(conc), 180 °C	No reaction	$C_6H_4(NH_2)SO_2OH$ 4-aminobenzenesulphonic acid
HNO_3/H_2SO_4	Chars	Oxidizes and chars. No nitration

Diazotization

Occurs when a primary aromatic amine reacts with HNO_2 below 10 °C. Benzenediazonium ion is $C_6H_5{-}N{=}N^+$. Benzenediazonium compounds form (a) phenols on warming, (b) C_6H_5Cl with HCl(conc) and CuCl, (c) C_6H_5Br with HBr(conc) and CuBr, and (d) C_6H_5I with KI(aq). They couple with phenols and amines to form bright dyes.

QUESTIONS

1. Compare the properties of ammonia and methylamine. Outline two methods of preparation of methylamine.

2. Explain the meanings of the terms ***primary*** and ***secondary*** as applied to amines and to alcohols. Why is the compound $CH_3CH_2CHNH_2CH_3$ classified as primary, but the compound $CH_3CH_2CHOHCH_3$ classified as secondary? Write the displayed formulae of a primary, secondary and tertiary amine of molecular formula $C_4H_{11}N$, and name the compounds.

3. Describe in detail the laboratory preparation of phenylamine (aniline) from nitrobenzene.

4. Compare and contrast the physical properties of ethylamine and phenylamine (aniline). What is the action of these two amines on (a) water, (b) hydrochloric acid, (c) ethanoyl chloride?

5. Name the compound of formula $CH_3CONHC_6H_5$, and outline one method of preparation from nitrobenzene. How would you obtain 4-nitrophenylamine from the named compound? Why is it not possible to obtain the 4-nitro-compound by direct nitration of phenylamine?

6. What is meant by ***diazotization***, and under what conditions is it carried out? Explain how benzenediazonium chloride may be converted to one azo-compound.

7. An aromatic compound contained 79.3 per cent of carbon, 9.09 per cent of hydrogen, and nitrogen only. It formed a salt with concentrated hydrochloric acid, but the compound itself was only feebly basic. The compound did not react with nitrous acid, HNO_2. Show that the formula of the compound is probably $C_6H_5N(CH_3)_2$.

8. (a) A hydrocarbon contained 82.76 per cent by mass of carbon. Find its molecular formula and write structures for possible isomers. Suggest how the structures of the isomers might be decided. (b) Benzenediazonium chloride (molecular formula $N_2C_6H_5Cl$) reacts as follows: (i) a solution of it gives an instantaneous precipitate of silver chloride with silver nitrate; (ii) when boiled

with water, phenol is formed; (iii) when it is reduced, it gives phenylhydrazine. Interpret these reactions and show how the structure of benzenediazonium chloride may be deduced from them. (C.)

9. From combustion of 1 g of a liquid organic compound, ***X***, 2.84 g of carbon dioxide, 0.68 g of water, and 120.4 cm^3 of nitrogen (measured at s.t.p.) are obtained. At 546 K and 760 mmHg pressure, 1000 cm^3 of the vapour of ***X*** weighs 2.08 g. Find the molecular formula of ***X*** and suggest a structural formula for it. How would you expect ***X*** to react with (a) bromine, (b) nitrous acid, HNO_2? (C.)

10. Describe and explain chemical tests, one in each case, by which you could distinguish between (a) ethylamine and phenylamine, (b) ethanol and phenol, (c) chlorobenzene and trichloromethane, (d) methanol and ethanol, (e) ethane and ethene. (C.)

11. The amino-acid of structure CH_3CHNH_2COOH can be prepared from 2-bromopropanoic acid by reaction with large excess of aqueous ammonia. Why is it advisable to use a large excess of the reagent? Write an equation for the reaction and suggest a possible mechanism. The aqueous amino-acid is electrolysed at pH = 1 and then pH = 13, and the amino-acid migrates to the cathode and the anode respectively. Write probable structures for the migratory species in each case. Samples of the natural and the synthetic amino-acid do not behave in the same way when examined in plane polarized light. State the difference in behaviour and account for the difference.

12. Phenylamine is prepared by reducing nitrobenzene with tin and hydrochloric acid. The mixture is then made alkaline with sodium hydroxide and steam distilled. The distillate is extracted with ether and the ethereal solution is allowed to stand over solid sodium hydroxide. The ethereal solution is then removed and distilled to obtain the amine, which is then distilled again at a higher temperature. (a) What is the function of the first alkali added and of the solid sodium hydroxide? (b) Why is steam distillation successful in this preparation? (c) Why is it relatively easy to separate the ether and amine present in the ethereal solution? (d) How and under what conditions does phenylamine react with aqueous sodium nitrite, $NaNO_2$, in the presence of dilute hydrochloric acid?

13. The structural formula of compound ***A*** is $CH_3CH_2CH_2COOCH(CH_3)_2$. When heated with aqueous ammonia it gives two products, ***B***, C_4H_9NO, and ***F***, C_3H_8O. ***B*** can be converted to ***C***, C_3H_9N, which reacts with aqueous $NaNO_2$ and acid to yield compound ***D*** isomeric with ***F***. ***D*** and ***F*** give isomeric compounds ***E*** and ***G***, C_3H_6O, on treatment with acidified potassium dichromate solution. Both ***E*** and ***G*** give precipitates with hydroxylamine. ***G*** has no reaction with Tollen's reagent whereas ***E*** produces a mirror on the side of the tube. Name and give the structural formulae for compounds ***B*** to ***G***, and outline a method by which ***B*** may be converted to ***C***. (C.)

14. Describe and briefly explain what happens in each of the following experiments and write balanced equations for the reactions that occur. (a) Propan-2-ol is warmed with acidified aqueous potassium dichromate(VI). (b) Aqueous bromine is added to aqueous phenol. (c) Phenylethene (a liquid) is allowed to stand in the air for some time. (d) Cold, aqueous sodium nitrite is slowly added to a solution of phenylamine in hydrochloric acid at 5°C and the resulting mixture is poured into an alkaline solution of phenol. (C.)

Chapter 20
AMIDES

Amides may be regarded as derivatives of ammonia in which one hydrogen atom is replaced by an RCO group (R = H or alkyl or aryl group). Their general

formula is $RCONH_2$ or $R-\underset{\underset{O}{\|}}{C}-NH_2$ and their functional group is $-C\begin{smallmatrix}\nearrow O \\ \searrow NH_2\end{smallmatrix}$

Amides may also be regarded as derivatives of acids in which the OH group is replaced by the amino group NH_2.

$HCONH_2$	CH_3CONH_2	$C_6H_5CONH_2$
methanamide	ethanamide	benzamide

Preparation of amides

1. *Dehydration of ammonium salts of carboxylic acids.* Mix an ammonium salt with an equal mass of its own anhydrous acid (e.g. ammonium ethanoate and glacial ethanoic acid), heat under reflux for 2 hours, and then distil:

$$CH_3COO^-NH_4^+ \xrightarrow{\text{dehydration}} H_2O + CH_3CONH_2$$

The acid prevents dissociation of the salt into ammonia and acid by displacing the following equilibrium to the left:

$$CH_3COONH_4 \rightleftharpoons CH_3COOH + NH_3$$

2. *Hydrolysis of a nitrile.* Dissolve a nitrile in concentrated sulphuric acid and then add the mixture to water:

$$RC{\equiv}N + H_2O \rightarrow RCONH_2$$

Hydrolysis with water alone is too slow; hydrolysis in the usual way with acids and alkalis converts the nitrile to an ammonium salt. This method has almost no practical value.

3. *Action of ammonia on acid chlorides and acid anhydrides.* Aliphatic acid chlorides react violently with ammonia solution at room temperature (p. 197); aromatic acid chlorides react more slowly; the actions of acid anhydrides and esters are much slower:

$$CH_3COCl + 2NH_3(aq) \rightarrow NH_4Cl + \underset{\text{ethanamide}}{CH_3CONH_2}$$

$$C_6H_5COCl + 2NH_3(aq) \rightarrow NH_4Cl + \underset{\text{benzamide}}{C_6H_5CONH_2}$$

$$(CH_3CO)_2O + 2NH_3(aq) \rightarrow CH_3COONH_4 + CH_3CONH_2$$

Tests on ethanamide

1. *Solubility and basic properties.* Add water to a few crystals of ethanamide. Note if the amide is soluble. Test the mixture with litmus.

2. *Hydrolysis.* Add about 5 cm^3 of dilute sodium hydroxide to ethanamide crystals and warm. Identify the gas evolved by its smell and its action on litmus paper. Repeat with dilute sulphuric acid instead of alkali.

3. *Nitrous acid, HNO_2.* Dissolve a few crystals of sodium nitrite in the minimum volume of cold water and add an equal volume of dilute sulphuric or hydrochloric acid. Add ethanamide crystals and observe any reaction. Neutralize the liquid by dilute sodium hydroxide and do the iron(III) chloride test (p. 179).

Properties of amides

Physical properties. Methanamide is a liquid at room temperature. Ethanamide and other amides are white crystalline solids. Ethanamide smells of mice, but the smell is due to impurities. The boiling points of amides are higher than those of the corresponding acids because the two hydrogen atoms in each $-CONH_2$ group form more hydrogen bonds than the $-COOH$ group in acids. The lower amides are soluble in water, ethanol and ether.

Chemical properties

1. *Basic properties.* Aqueous ethanamide is neutral. Ethanamide and other amides are very weak bases. The carbon–oxygen bond is polar: $\overset{\delta+}{C}=\overset{\delta-}{O}$. The electron-attracting power of the oxygen atom causes delocalization of electrons and inhibits the taking up of a proton; amides are therefore much less basic than amines:

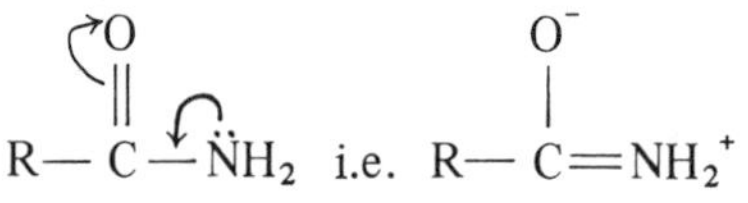

$$R-\overset{\overset{O}{\|}}{C}-\ddot{N}H_2 \quad \text{i.e.} \quad R-\overset{\overset{O^-}{|}}{C}=NH_2^+$$

Amides form salts only with strong acids:

$$CH_3CONH_2 + HCl \rightarrow CH_3CONH_3^+Cl^- \text{ (in dry ether)}$$

2. *Hydrolysis.* Amides hydrolyse readily when warmed with either aqueous sodium (or potassium) hydroxide or a dilute mineral acid:

$$CH_3CONH_2(aq) + OH^-(aq) \rightarrow CH_3COO^- + NH_3(g)$$

$$CH_3CONH_2(aq) + H_3O^+(aq) \rightarrow CH_3COOH + NH_4^+(aq)$$

3. *Dehydration.* Phosphorus(V) oxide, phosphorus trichloride oxide or sulphur dichloride oxide convert an amide to a nitrile on warming:

$$RCONH_2 \rightarrow RCN + H_2O$$

4. *Nitrous acid.* Nitrogen is evolved and acid is formed.

$$\begin{matrix} CH_3CO\,N\,H_2(s) \\ + \\ HO\,N\,O(aq) \end{matrix} \rightarrow CH_3COOH(aq) + N_2(g) + H_2O$$

5. *Reduction.* Lithium tetrahydridoaluminate(III) reduces an amide to a primary amine:

$$CH_3CONH_2 \rightarrow CH_3CH_2NH_2$$

SUMMARY

Amides

$RCONH_2$ (R = H, alkyl or aryl). Functional group: $-C\begin{matrix}\nearrow O \\ \searrow NH_2\end{matrix}$

Preparation

1. Reflux an ammonium salt with its carboxylic acid.
2. Hydrolyse a nitrile with acid or alkali.
3. Acid chloride or anhydride with ammonia.

Properties

1. Weak bases. Neutral aqueous solutions. Salts with strong acids.
2. Acids or alkalis hydrolyse to a carboxylic acid.
3. P_2O_5 or SCl_2O dehydrate to a nitrile.
4. Nitrous acid forms nitrogen and a carboxylic acid.
5. $Li[AlH_4]$ (ether) reduces to a primary amine.

QUESTIONS

1. Outline two methods of preparing ethanamide. How does this amide react with (a) dilute sulphuric acid, (b) dilute sodium hydroxide, (c) phosphorus(V) oxide, and (d) ethanoyl chloride?

2. By reference to ammonium ethanoate, ethanamide and ethanenitrile, show that an amide may be regarded as intermediate between an ammonium salt and a nitrile. Indicate briefly how each of the three named compounds can be converted into the other two.

3. How, and under what conditions, does potassium hydroxide react with the following: (a) ethanamide, (b) benzenesulphonic acid, (c) bromoethane, (d) ethanal, and (e) trichloromethane? For each of (a), (b) and (c) outline a method by which you would isolate from the reaction mixture one pure compound which contains only carbon, hydrogen and oxygen. (C.)

4. Ozonolysis of compound *A*($C_{11}H_{14}$) gave *B*(C_8H_8O) and *C*(C_3H_6O) in equimolar amounts. *C* was unaffected by ammoniacal silver nitrate, but *B* reacted with this reagent to give *D* ($C_8H_8O_2$). When *D* was treated successively with phosphorus trichloride and ammonia, *E* (C_8H_9NO) was formed. *E* reacted with bromine and sodium hydroxide to give *F* (C_7H_9N), a much weaker base than methylamine. Identify the compounds *A*–*F* as far as possible and comment on the reactions. (C.)

5. An organic compound *A* contains 23.7 per cent by mass of nitrogen. *A* when refluxed with sodium hydroxide solution yields ammonia, and it also reacts with nitrous acid, HNO_2, giving nitrogen. Distillation of a mixture of *A* with phosphorus(V) oxide gives a compound *B*, which is composed of carbon, hydrogen

and nitrogen, the percentage of nitrogen being 34.14. Calculate the mass of ***B*** which contains 14 g of nitrogen. Thus identify ***B*** and ***A***, and account for the above reactions. (C.)

6. Give one example of the use of each of the following reagents in organic chemistry: (a) lithium tetrahydridoaluminate(III), $Li[AlH_4]$, (b) potassium cyanide, (c) sulphur dichloride oxide, SCl_2O, (d) potassium manganate(VII), $KMnO_4$, and (e) tin.

7. Write equations and give reagents and conditions to show how, given supplies of benzene and methylbenzene, you would prepare the following compounds: (a) C_6H_5I, (b) $C_6H_5NHCOCH_3$, (c) C_6H_5CHO. An aromatic compound is believed to have the structure $NH_2C_6H_4OCH_2COOC_2H_5$. By what chemical reactions would you show the presence of (d) the benzene ring, (e) the amino group, (f) the ester group? (C.)

8. Substituted aromatic compounds are often synthesized by nitrating the parent hydrocarbon followed by further reactions with the product. Illustrate this statement by discussing the synthesis of the following from either benzene or methylbenzene: (a) 3-nitrophenylamine, (b) 2-aminomethylbenzene, (c) 4-nitrobenzoic acid, (d) phenylethanenitrile. Give the names and formulae of all reagents and the structures of the intermediate compounds and the principal by-products (where appropriate). Specify the reaction conditions that ensure high yields of the correct products.

9. Identify the compounds ***A, B,*** and ***C*** from their following reactions. Explain your reasoning and write suitable equations where possible. (i) ***A***, C_2H_5NO, is neutral. It does not react with cold aqueous sodium hydroxide but slowly forms ammonia with the hot alkali. (ii) ***B***, $C_3H_3NO_2$, is acidic, and slowly forms ammonia on refluxing with aqueous sodium hydroxide. (iii) ***C***, C_8H_9NO, is neutral. Concentrated aqueous sodium hydroxide hydrolyses it to a salt, $C_2H_3O_2Na$, and a compound C_6H_7N, which forms a compound $C_6H_4Br_3N$ with aqueous bromine.

10. For each of the following pairs of isomers describe one chemical test to distinguish between them. State clearly the conditions of the reactions, the observations made, and write equations. (i) $CH_3CH_2CH_2NH_2$ and $(CH_3)_3N$, (ii) $CH_2{=}CHCH_2CONH_2$ and $CH_3CH_2CHOHCN$, (iii) CH_3COOH and $CHO{-}CH_2OH$.

11. The reactivity of the amino ($-NH_2$) group is largely dependent on the nature of the group to which it is attached. Give examples which illustrate the difference in reactivity of the amino group when attached to (a) an alkyl group; (b) an aromatic nucleus; (c) a carbonyl group. How may this difference in reactivity be explained? (L.)

Chapter 21
PROTEINS

AMINO-ACIDS

These compounds contain two functional groups in the same molecule – the acidic carboxyl group COOH and the basic amino group NH_2. They can react as acids and as bases. The general formula is $NH_2{-}CHR{-}COOH$ (R = H, alkyl or other group).

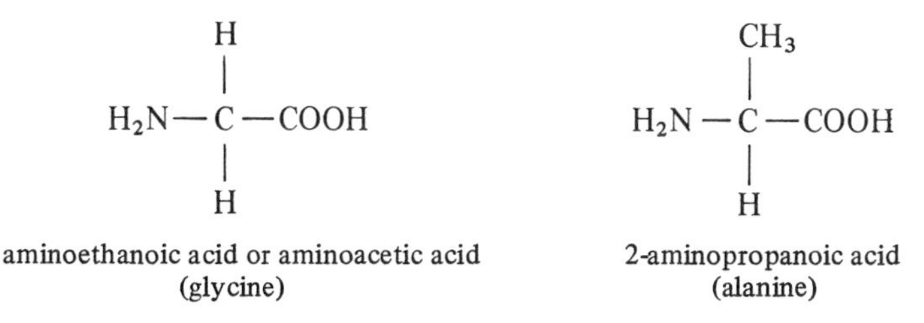

aminoethanoic acid or aminoacetic acid (glycine)

2-aminopropanoic acid (alanine)

All amino-acids obtained from naturally occurring compounds have the NH_2 group joined to the carbon atom next to the COOH group, i.e. they are 2-amino-acids. All amino-acids except aminoethanoic acid have four different groups joined to the second carbon atom, which is a chiral carbon atom (p. 205). Therefore optical isomers exist.

Aminoethanoic acid, NH_2CH_2COOH

The acid is prepared by adding aqueous chloroethanoic acid to excess concentrated ammonia solution and then boiling to expel excess ammonia:

$$2NH_3(aq) + ClCH_2COOH \rightarrow NH_4Cl + NH_2CH_2COOH$$

Tests on aminoethanoic acid.

1. *Solubility and acidity or basic properties.* Add water to the acid. Note if it is soluble. Test the mixture with neutral litmus and universal indicator. Find if the acid is soluble in ethanol and in ether.

2. *Nitrous acid.* To an aqueous solution of the amino-acid, add sodium nitrite, $NaNO_2$, and dilute hydrochloric acid. Is a gas evolved and, if so, is it nitrogen?

3. *Salt formation.* To separate aqueous solutions of the acid add a few drops of aqueous copper(II) sulphate and iron(III) chloride. Note any colours or precipitates.

Properties of aminoethanoic acid

Physical properties. The acid is a white solid (m.p. 507 K, b.p. 559 K) with a sweet taste. It is soluble in water, forming a neutral solution, and only slightly soluble in ethanol and ether.

Chemical properties. Both in the solid state and in solution the molecules exist as *zwitterions* (which means 'mongrel ions'), formed by the transfer of a proton from the carboxyl group to the amino group: $H_3\overset{+}{N}CH_2COO^-$. Two facts indicate the existence of zwitterions: (a) the high melting point of the acid shows that the molecules are highly polar and attract each other strongly, just like the ions of ionic compounds, and (b) the acid forms a trimethyl derivative with all three methyl groups attached to the nitrogen atom when the acid is methylated with iodomethane:

$$3CH_3I + H_3\overset{+}{N}CH_2COO^- \rightarrow (CH_3)_3\overset{+}{N}CH_2COO^- + 3HI$$

1. *Acidic properties.*

(a) The acid forms salts with alkalis, e.g. NH_2CH_2COONa, and a blue copper(II) salt, $(H_2NCH_2COO)_2Cu{\cdot}H_2O$.

(b) It forms esters with alcohols in the presence of an acid catalyst:

$$H_2NCH_2COOH + C_2H_5OH \rightarrow \underset{\text{ethyl aminoethanoate}}{H_2NCH_2COOC_2H_5} + H_2O$$

(c) It loses carbon dioxide (decarboxylation) when heated with sodalime or calcium oxide:

$$H_3\overset{+}{N}CH_2COO^-(s) + CaO(s) \rightarrow CaCO_3(s) + CH_3NH_2(g)$$

(d) It forms an amide, $H_2NCH_2CONH_2$.

2. *Basic properties.* It forms salts with acids, such as the crystalline $[H_3\overset{+}{N}CH_2COOH]Cl^-$ with hydrochloric acid.

3. *Nitrous acid.* The NH_2 group is replaced by OH:

$$H_2NCH_2COOH + HNO_2 \rightarrow N_2(g) + H_2O + \underset{\text{hydroxyethanoic acid}}{HOCH_2COOH}$$

4. *Acid chlorides and anhydrides.* Ethanoyl chloride, ethanoic anhydride and benzoyl chloride react and replace one of the hydrogen atoms of the NH_2 group:

$$CH_3COCl + H_2NCH_2COOH \rightarrow HCl + \underset{\text{(ethanoylamino)ethanoic acid}}{CH_3CONHCH_2COOH}$$

PROTEINS

Proteins are formed in most animal and vegetable tissues. They contain nitrogen, and some also contain sulphur and phosphorus. Examples are:

haemoglobin (in blood)	casein (in milk)
keratin (in hair, horn, feathers)	albumen (in eggs)
gelatin (in bones)	insulin (in animal pancreas)

Enzymes (e.g. pepsin, rennin, trypsin) are proteins, and life depends on the biochemical reactions which they catalyse.

The relative molecular masses of proteins are high, varying between several thousands and a million or more. Those of insulin, albumen and haemoglobin are 6000, about 50 000, and about 70 000 respectively. The formula of insulin, the simplest protein, is $C_{254}H_{377}N_{65}O_{75}S_6$. The molecules are either linear (linked by hydrogen atoms into long, twisted rods) or three-dimensional and globular.

Tests on proteins

1. *Heat.* Heat egg white and yolk separately and observe any change. (Egg protein has globular molecules.)

2. *Sodalime.* Heat a dried protein with sodalime and identify the gas evolved. (This test shows that proteins contain nitrogen.)

Properties of proteins

1. *Hydrolysis.* A protein molecule contains at least 50 peptide groups,

$$-CO-NH- \text{ or } \begin{array}{c} -C-N- \\ \| \quad | \\ O \quad H \end{array}$$

which link together the amino-acid units.

Hydrolysis breaks these groups:

$$-CO-NH- + H_2O \rightarrow -COOH + -NH_2$$

Partial hydrolysis breaks some but not all of the groups and produces smaller molecules of substances called *polypeptides* and then even smaller molecules of *peptides.* Complete hydrolysis produces a mixture of amino-acids:

$$\text{Proteins} \rightarrow \text{Polypeptides} \rightarrow \text{Peptides} \rightarrow \text{Amino-acids}$$

2. *Heat.* Soluble proteins form colloidal solutions, which coagulate on heating and form insoluble solids.

3. *Sodalime.* Ammonia is evolved.

Structure of proteins

A mixture of amino-acids forms when a protein is hydrolysed for 24 hours with 6 M hydrochloric acid. About 30 acids have been identified, and 9 are essential to man. Insulin forms a mixture of 17 amino-acids and its molecule contains 51 amino-acid units or residues; a haemoglobin molecule contains 574 amino-acid units.

A protein is probably formed by condensation polymerization of amino-acids, with elimination of water. The simplest possible example of this polymerization is:

$$H_2NCH_2COOH + H_2NCH_2COOH \rightarrow H_2O + H_2N-CH-\underbrace{CO-NH}_{\text{peptide link}}-CH-COOH$$

All proteins contain many (at least 50) peptide links, CONH, in their molecules and they are *polyamides.* The four atoms of a peptide link and the two carbon atoms to which they are joined are all in one plane. The different amino-acids can be used in various proportions and they can be arranged in the polymeric protein system in various orders. Therefore an infinite number of different protein structures is possible.

The C=O and N—H groups in a peptide link are polar and form hydrogen bonds. Each protein molecule has many hydrogen bonds:

$$>\overset{\delta+}{C}=\overset{\delta-}{O} \cdots\cdots \overset{\delta+}{H}-\overset{\delta-}{N}<$$

The nature and number of amino-acid units present in a protein is determined by hydrolysis, and the amino-acids are separated and identified by chromatography. Partial hydrolysis of a protein, using certain enzymes, produces peptides of moderate chain lengths; identification of the peptides enables the amino-acid sequence in the protein to be deduced.

X-ray diffraction is used to determine the shape in space (the *conformation*) of protein molecules. Some protein molecules are shaped like long rods, and the peptide groups occur at regular intervals. The rods may be twisted into helical (spiral) shapes. Other protein molecules are globular because the peptide groups are rolled up. *Fibrous* proteins (e.g. hair) have rod-like molecules, and they are insoluble in water; *globular* proteins (e.g. egg white, milk) have globular molecules, and they are soluble in water or acids or bases. Hydrogen bonds enable the protein molecules to keep their helical or globular shapes.

NYLON

Nylon is made from hexane-1,6-diamine, $H_2N(CH_2)_6NH_2$, and hexane-1,6-dioic acid, $HOOC(CH_2)_4COOH$. Since each molecule contains 6 carbon atoms, the nylon formed is called nylon 66. Water is eliminated between an amino group and a carboxyl group, and nylon is a *polyamide* and a condensation polymer:

$$nHO_2C{-}(CH_2)_4{-}CO_2H + nH_2N{-}(CH_2)_6{-}NH_2$$

$$\rightarrow \;\left[OC{-}(CH_2)_4{-}CO{-}NH{-}(CH_2)_6{-}NH\right]_n + (n-1)H_2O$$

n is at least 40, and the relative molecular mass is 8000 or more.

A simple equation showing only two molecules combining is:

$$HO{-}CO{-}X{-}CO{-}OH + H_2N{-}Y{-}NH_2$$

$$\rightarrow HO{-}CO{-}X{-}CO{-}X{-}CO{-}NH{-}Y{-}NH_2 + H_2O$$

$$X = (CH_2)_4 \qquad\qquad Y = (CH_2)_6$$

In the polymer molecule the repeat unit is:

$$-CO{-}X{-}CO{-}NH{-}Y{-}NH-$$

and since it contains many amide groups (—CO—NH—) nylon 66 is a polyamide.

A simple block diagram for nylon is:

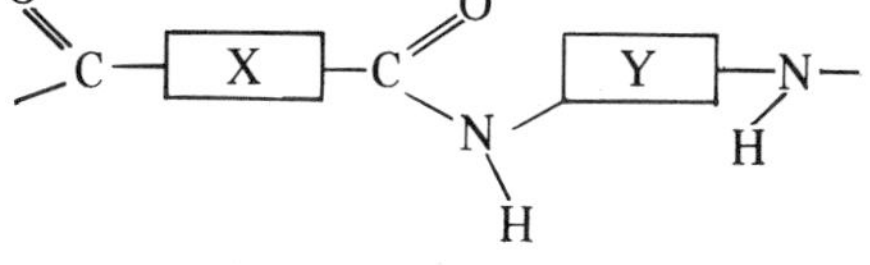

O
C—[X]—C=O
N—[Y]—N—
H H

The —CO—NH— groups in nylon form hydrogen bonds between different molecules and make nylon fibres and sheets very strong. Most of a nylon macromolecule consists of $(CH_2)_4$ and $(CH_2)_6$ groups. These long hydrocarbon chains ensure that the substance does not absorb water, whereas cotton with many OH groups and wool with many N—H and C=O groups do absorb water.

Nylon molecules are long and unbranched and have no bulky side-groups. If solid nylon is stretched during its manufacture, a process called *cold drawing*, the long molecules are arranged in the same direction and the substance becomes more crystalline, stronger and more pliable. Drawn nylon fibres are continuous, straight, smooth and lustrous.

Uses of nylon. Nylon is light and elastic, does not absorb water, is resistant to acids, does not fade, keeps its shape, is insect-proof, can be dyed, and readily sheds unwanted creases, but keeps permanent creases. It is therefore used for clothing (suits, shirts, stockings, socks, underwear), carpets and curtains, camping equipment, parachutes, and so on. Because sunlight and water do not affect it, nylon is very useful in kitchens, bathrooms, and even around swimming pools.

SUMMARY

Amino-acids

$H_2N-CHR-COOH$ (R = H, alkyl or other group). Functional groups: $-NH_2$ and $-COOH$.

Properties of aminoethanoic acid

1. Acidic properties. Forms salts with alkalis, esters with alcohols, and methylamine with sodalime.

2. Amine properties. Forms salts with acids, hydroxyethanoic acid and nitrogen with nitrous acid, and ethanoyl and benzoyl derivatives with acid chlorides and anhydrides.

The aqueous solution contains zwitterions: $H_3\overset{+}{N}CH_2COO^-$.

Proteins

Formed by condensation polymerization of amino-acids. Contain peptide linkages, —CONH—. Hydrolysis breaks the linkages and forms amino-acids. Heat coagulates. Hot sodalime forms ammonia.

Nylon

Nylon 66 is manufactured from hexane-1,6-diamine, $H_2N(CH_2)_6NH_2$, and hexane-1, 6-dioic acid, $HOOC(CH_2)_4COOH$.

QUESTIONS

1. How is chloroethanoic acid prepared from ethanoic (acetic) acid. What is the action of ammonia on this chloro-acid? Mention two chemical reactions of the product.

2. All proteins are polyamides, and nylon is also a polyamide. Explain fully the meaning of this statement.

3. An amino-acid of molecular formula $C_3H_7O_2N$ combines with hydrochloric acid. The compound formed contains chlorine. Show that the percentage composition by mass of chlorine in this compound is 28.3 per cent. Write an equation for the reaction.

4. Describe how you would distinguish between the following pairs of isomers by two simple chemical tests in each case. Give equations for the reactions involved.

(a) $CH_2{=}CH{-}CH{=}CH_2$ and $CH_3CH_2C{\equiv}CH$
(b) $CH_3CH_2CH_2OH$ and $(CH_3)_2CHOH$
(c) $CH_3C_6H_4NH_2$ and $C_6H_5CH_2NH_2$
(d) $HOCH_2CONH_2$ and NH_2CH_2COOH. (C.)

5. Name (a) one commercial plastic formed by addition polymerization, (b) one commercial plastic formed by condensation polymerization, (c) one naturally occurring polymer. Give the monomers from which these three polymers are formed and give an account of the new chemical bonds formed in polymerization and of the conditions necessary for the formation of the polymers named in (a) and (b). What features of molecular structure should be present in polymers which are required to be (i) elastic, (ii) rigid? (C.)

6. An organic compound ***A*** contains 18.67 per cent of nitrogen by mass. When treated with sodium hydroxide, ***A*** forms a salt but ammonia is not produced. ***A*** reacts with nitrous acid, HNO_2, with the evolution of nitrogen, forming an acid ***B***, which contains 63.15 per cent of oxygen. On oxidation, ***B*** gives an acid ***C*** which (a) decolorizes a hot solution of potassium manganate(VII), $KMnO_4$, acidified with dilute sulphuric acid, (b) after neutralizing with ammonia solution, forms a white precipitate with calcium chloride. Calculate the mass of ***A*** which contains 14 g of nitrogen. Identify ***A, B*** and ***C*** giving your reasoning, and account for the above reactions. (C.)

7. (a) How, and under what conditions, does nitrous acid react with (i) ethylamine, (ii) phenylamine? Outline the industrial importance of the reaction of an aromatic amine with nitrous acid. (b) What functional groups are present in the molecule of an amino-acid? Aminoethanoic acid is a high melting point solid which is soluble in water giving a neutral solution, but insoluble in ethoxyethane ('ether'). Draw a full structural formula for aminoethanoic acid which is consistent with these observations and explain how you arrive at your answer. (C.)

8. Write short explanatory notes of each of the following: (a) The structures of proteins. (b) The increase in acid strength as the hydrogen atoms of the $-CH_3$ group in ethanoic acid are successively substituted by chlorine atoms. (c) The mechanism of one addition reaction (of your own choice) in organic chemistry. (L.)

APPENDIX

IMPORTANT REAGENTS AND THEIR REACTIONS

Listed below are the more useful reagents in organic chemistry, together with their reactions and examples of these. The numbers are those of the pages on which the reactions are described. The reagents are in alphabetical order by name, not symbol.

Reagent	*Reaction*	*Examples*
KOH(aq), NaOH(aq) i.e. alkali, OH^-(aq)	Hydrolysis	$C_2H_5Br \rightarrow C_2H_5OH$ 96 $C_6H_5CH_2Cl \rightarrow C_6H_5CH_2OH$ 107 $CH_2BrCH_2Br \rightarrow CH_2OHCH_2OH$ 99 C_6H_5Br no reaction 104 $CH_3COOC_2H_5 \rightarrow CH_3COOH + C_2H_5OH$ 213 $CH_3CN \rightarrow CH_3COOH$ 223 $C_6H_5CN \rightarrow C_6H_5COOH$ 223 $CH_3CONH_2 \rightarrow CH_3COOH$ 246 Fat or oil → soap + glycerol 216
	Cannizzaro	$C_6H_5CHO \rightarrow C_6H_5COOH + C_6H_5CH_2OH$ 127, 153
Al/Hg + alcohol, Zn/Cu + alcohol	Reduction of halogenoalkanes	$C_2H_5I \rightarrow C_2H_6$ 19
$AlCl_3$	Friedel-Crafts (alkylation and acylation)	$C_6H_6 + ClCH_3 \rightarrow C_6H_5CH_3$ 79 $C_6H_6 + CH_3COCl \rightarrow C_6H_5COCH_3$ 79
$AlCl_3$, $FeCl_3$, $FeBr_3$ (halogen carriers)	Halogenation	$C_6H_6 + Cl_2 \rightarrow C_6H_5Cl$ 77 $C_6H_6 + Br_2 \rightarrow C_6H_5Br$ 77 $C_6H_5CH_3 + Cl_2 \rightarrow C_6H_4ClCH_3$ 84

Reagent	*Reaction*	*Examples*
Al_2O_3	Dehydration Catalytic cracking	$C_2H_5OH \rightarrow CH_2{=}CH_2$ 41 Petroleum alkanes to aromatic and branched-chain hydrocarbons 71
C_6H_5COCl	Benzoylation 198	$C_6H_5OH \rightarrow C_6H_5COOC_6H_5$ 197 $C_6H_5NH_2 \rightarrow C_6H_5CONHC_6H_5$ 197 $NH_3(aq) \rightarrow C_6H_5CONH_2$ 197 NH_2CH_2COOH $\rightarrow C_6H_4CONHCH_2COOH$ 251
CH_3COCl, $(CH_3CO)_2O$	Ethanoylation (acetylation) 198	$NH_3(aq) \rightarrow CH_3CONH_2$ 197 $C_2H_5NH_2 \rightarrow CH_3CONHC_2H_5$ 197 NH_2CH_2COOH $\rightarrow CH_3CONHCH_2COOH$ 251 $C_2H_5OH \rightarrow CH_3COOC_2H_5$ 197 $C_6H_5OH \rightarrow CH_3COOC_6H_5$ 130
Fehling's soln. (Cu^{2+} complex ion)	Test for aliphatic aldehydes	Red precipitate: Cu_2O 153 Not with benzaldehyde or ketones 153 Glucose and fructose react 170
I_2 + alkali Br_2 + alkali	Trihalomethane (haloform) test for CH_3CHOH- or CH_3CO- 158	$CH_3CH_2OH \rightarrow CHI_3$, 115 yellow solid $CH_3CH_2OH \rightarrow CHBr_3$ 115 $CH_3COCH_3 \rightarrow CHI_3$ 158 $CH_3COC_2H_5 \rightarrow CHI_3$ 158
Halogens (Cl_2, Br_2)	Chlorination Bromination	$CH_4 \rightarrow CH_3Cl$, CH_3Cl, CH_2Cl_2, CCl_4 31 $C_6H_6 \rightarrow C_6H_6Cl_2 \rightarrow C_6H_6Cl_4$ $\rightarrow C_6H_6Cl_6$ 76 $C_6H_5CH_3 \rightarrow C_6H_5CH_2Cl$ $\rightarrow C_6H_5CHCl_2 \rightarrow C_6H_5CCl_3$ 84 $C_6H_5CH_3 \rightarrow C_6H_4ClCH_3$ 84 (halogen-carrier) $C_6H_5OH \rightarrow C_6H_2Cl_3OH$ 131 $C_6H_5OH \rightarrow C_6H_2Br_3OH(s)$, 131 white $C_6H_5NH_2 \rightarrow C_6H_2Br_3NH_2(s)$, 238 white $CH_3COOH \rightarrow CH_2ClCOOH$ $\rightarrow CHCl_2COOH$ $\rightarrow CCl_3COOH$ 183

Reagent	*Reaction*	*Examples*
Br_2(aq)	Bromination	Test for unsaturation, i.e. C=C or C≡C 51, 46, 64
H_2(g)/Ni catalyst	Reduction of C=C, C≡C and C≡N	$CH_2{=}CH_2 \rightarrow CH_3CH_3$ 46, 51 $C_6H_6 \rightarrow C_6H_8 \rightarrow C_6H_{10} \rightarrow C_6H_{12}$ 76 $C_6H_5CH_3 \rightarrow C_6H_{11}CH_3$ 83 $C_6H_5OH \rightarrow C_6H_{11}OH$ 132 Oils → fats 215 $RCN \rightarrow RCH_2NH_2$ 223 $CH{\equiv}CH \rightarrow CH_3CH_3$ 63
H^+(aq)	Hydrolysis of esters, nitriles and amides	$CH_3COOC_2H_5 \rightarrow CH_3COOH + C_2H_5OH$ 213 $CH_3CN \rightarrow CH_3COOH + NH_4^+$ 223 $CH_3CONH_2 \rightarrow CH_3COOH + NH_4^+$ 246
H_2(g)/Pt	Reduction of C=C and C≡C	Alkene → alkane 46, 51 Alkyne → alkane 63 Platforming petroleum alkanes 71
Li[AlH_4] (ether), Na[BH_4] (aq)	Reduction of C=O 154	Ester → alcohol 213 Nitrile → amine 223 Amide → amine 246
HNO_3(dilute)	Oxidation	$C_6H_5CH_3 \rightarrow C_6H_5COOH$ 83
HNO_3(conc)/ H_2SO_4(conc)	Nitration	$C_6H_6 \rightarrow C_6H_5NO_2$ 78 $C_6H_5CH_3 \rightarrow CH_3C_6H_4NO_2 \rightarrow CH_3C_6H_2(NO_2)_3$ 85 $C_6H_5OH \rightarrow C_6H_2(NO_2)_3OH$ 132
HNO_3(conc)	Oxidation	Estimation of halogens 2 Estimation of sulphur 3
HNO_2 i.e. $NaNO_2/H^+$(aq)	Diazotization (of aromatic amines)	$C_6H_5NH_2 \rightarrow C_6H_5{-}N{=}N^+$ (5 °C) $\rightarrow C_6H_5OH + N_2$ 238 (above 10 °C) $CH_3CONH_2 \rightarrow CH_3COOH + N_2$ 246 $NH_2CH_2COOH \rightarrow HOCH_2COOH + N_2$ 251

Reagent	*Reaction*	*Examples*
O_3 (in hexane or other inert solvent)	Ozonolysis (addition to C=C bond)	$C_2H_4 \rightarrow C_2H_4O_3$ 49 $C_6H_6 \rightarrow C_6H_6(O_3)_3(s)$ 76
H_3PO_4	Dehydration	CH_3CH_2OH 200 °C $\rightarrow CH_2{=}CH_2$ 44
PCl_5, SCl_2O, PBr_3 ($P + Br_2$), $P + I_2$	Substitution of —OH by —Hal or of O in C=O by Cl_2 or Br_2	$C_2H_5OH \rightarrow C_2H_5Cl$ or C_2H_5Br or C_2H_5I 91 $CH_3COOH \rightarrow CH_3COCl(l)$ 182 $C_6H_5COOH \rightarrow C_6H_5COCl(l)$ 183 $CH_3CHO \rightarrow CH_3CHCl_2(l)$ 158 $(CH_3)_2CO \rightarrow (CH_3)_2CCl_2$ 158 $C_6H_5CHO \rightarrow C_6H_5CHCl_2$ 158
P_2O_5	Dehydration	$CH_3CONH_2 \rightarrow CH_3CN(l)$ 222
$KMnO_4$ + acid, i.e. $MnO_4^-/H^+(aq)$	Oxidation of hydrocarbons, aldehydes and ketones	$CH_2{=}CH_2 \rightarrow CH_2OHCH_2OH$ 48 $CH{\equiv}CH \rightarrow (COOH)_2$ 64 $C_6H_5CH_3 \rightarrow C_6H_5COOH$ 83 (100 °C) $CH_3CHO \rightarrow CH_3COOH$ 153 $CH_3COCH_3 \rightarrow CH_3COOH$ 154
$K_2Cr_2O_7$ + acid, $Na_2Cr_2O_7$ + acid, i.e. $Cr_2O_7^{2-}/H^+(aq)$	Oxidation of alcohols	$CH_3CH_2OH \rightarrow CH_3CHO \rightarrow CH_3COOH$ 114 $(CH_3)_2CHOH \rightarrow (CH_3)_2CO \rightarrow CH_3COOH$ 122
Na(ether)	Wurtz reaction Fittig reaction	$C_2H_5I \rightarrow C_2H_5C_2H_5$ 19 $C_2H_5I + CH_3I \rightarrow C_2H_5CH_3$ or C_3H_8 20 $C_6H_5Br + CH_3Br \rightarrow C_6H_5CH_3$ 82
H_2SO_4(conc)	Salts or esters Dehydration	$C_6H_5NH_2 \rightarrow C_6H_5NH_3HSO_4$ 237 $C_2H_5OH \rightarrow C_2H_5HSO_4$ 114 $C_2H_5OH \rightarrow C_2H_5OC_2H_5$ (140 °C) 138 $\rightarrow CH_2{=}CH_2$ (170 °C) 43 $HCOOH \rightarrow CO$ 184 $(COOH)_2 \rightarrow CO$ 180

Reagent	*Reaction*	*Examples*
H_2SO_4(fuming) or H_2SO_4(conc)	Sulphonation	$C_6H_6 \rightarrow C_6H_5SO_2OH$ 77 $C_6H_5CH_3 \rightarrow CH_3C_6H_4SO_2OH$ 85 $C_6H_5OH \rightarrow HOC_6H_4SO_2OH$ 132 $C_6H_5NH_2 \rightarrow NH_2C_6H_4SO_2OH$ 238 Alkylbenzenes → detergents 217
	Oxidation	Estimation of nitrogen 2
Tollen's reagent $[Ag(NH_3)_2]^+(aq)$	Test for aldehydes, glucose, fructose	CH_3CHO, cold → Ag mirror 153 C_6H_5CHO, warm → Ag mirror 153 $C_6H_{12}O_6 \rightarrow$ Ag mirror 170 Ketones no reaction 153

Chemical Nomenclature

The Association for Science Education gives two recommended or systematic names for several compounds. For each pair given below, the first name is the one used in this book.

Alternative recommended names	*Formula*
Aminoethanoic acid (aminoacetic acid)	NH_2CH_2COOH
Benzaldehyde (benzenecarbaldehyde)	C_6H_5CHO
Benzonitrile (benzenecarbonitrile)	C_6H_5CN
Benzoyl chloride (benzenecarbonyl chloride)	C_6H_5COCl
Benzamide (benzenecarboxamide)	$C_6H_5CONH_2$
Benzoate (benzenecarboxylate)	C_6H_5COO-
Benzoic acid (benzenecarboxylic acid)	C_6H_5COOH
Carbamide (urea)	$CO(NH_2)_2$
Ethanoate (acetate)	CH_3COO-
Ethanoic acid (acetic acid)	CH_3COOH
Ethanoic anhydride (acetic anhydride)	$(CH_3CO)_2O$
Ethanenitrile (acetonitrile)	CH_3CN
Ethanoyl chloride (acetyl chloride)	CH_3COCl
Methanamide (formamide)	$HCONH_2$
Methanoic acid (formic acid)	$HCOOH$

Derivatives of the above compounds also have two recommended names, for example:

Chloroethanoic acid (chloroacetic acid)	$CH_2ClCOOH$
2-Methylbenzoic acid (2-methylbenzene-carboxylic acid)	$CH_3C_6H_4COOH$
Methanoate (formate)	$HCOO-$

ANSWERS TO NUMERICAL QUESTIONS

'The University of London School Examinations Board accepts no responsibility whatsoever for the accuracy or method of working in the answers given.'

Chapter 1 (p. 13): 4. C_3H_8. 5. 1250 cm^3. 7. C_6H_{12}.
Chapter 2 (p. 38): 3. 372 K. 5. About 138 K; about 218 K; $(CH_3)_2CHCH_3$, 2-methylpropane.
Chapter 5 (p. 87): 8. 5346 kJ mol^{-1}.
Chapter 6 (p. 101): 7. $X = (CH_3)_2CHCH_2I$; $Y = (CH_3)_2C{=}CH_2$; $Z = (CH_3)_2CICH_3$. 9. (a) Butan-2-ol; (b) $CH_2{=}CHCH_2CH_3$ (but-1-ene) and $CH_3CH{=}CHCH_3$ (but-2-ene); (c) (i) $CH_2BrCHBrCH_2CH_3$ and $CH_3CHBrCHBrCH_3$, (ii) $CH_3CHBrCH_2CH_3$ and $CH_3CH_2CHBrCH_3$. 10. (a) As in 9(a); (b) $CH_2{=}CHCHBrCH_3$ (3-bromobut-1-ene), $CH_3CBrCHCH_3$ (2-bromobut-2-ene), $CH_2{=}C{=}CHCH_3$ (buta-1,2-diene), $CH_2{=}CH{-}CH{=}CH_2$ (buta-1,3-diene), $CH_3C{\equiv}CCH_3$ (but-2-yne).
Chapter 7 (p. 108): 4. $C_6H_5CCl_3$.
Chapter 8 (p. 124): 7. CH_3O; $C_2H_6O_2$; $X = (CH_2OH)_2$.
Chapter 9 (p. 135): 6. C_7H_6O; 2-, 3- and 4-HOC_6H_4COOH.
Chapter 10 (p. 143): 4. $X = CH_3OC_3H_7$; $Y = CH_3I$; $Z = C_3H_7I$ 7. C_3H_8O; $A = CH_3OC_2H_5$; $B = CH_3CH_2CH_2OH$.
Chapter 11 (p. 164): 8. $A = CH_3CH_2CH{=}C(CH_3)_2$; $B = (CH_3)_2CO$; $C = CH_3CH_2CHO$; $D = CH_3CH_2CH_2OH$; $E = CH_3CH_2CH_2Br$; $F = CH_3CH{=}CH_2$. 9. $G = C_6H_5CH_2Cl$; $H = C_6H_5CH_2OH$; $J = C_6H_5CHO$; $K = C_6H_5CH{=}NNHC_6H_5$. 11. C_4H_8O; C_4H_8O; $CH_3COCH_2CH_3$, $CH_3CH_2CH_2CHO$, $(CH_3)_2CHCHO$; A = $HOCH_2COOH$, B = $CHOCOOH$, $C = (COOH)_2$.
Chapter 13 (p. 191): 7. Benzene, phenol, methanoic acid.
Chapter 15 (p. 207): 3. 2, 8, 16. 4. (b) +66°; (c) −5°. 7. $A = CH_3CH{=}CHCH_3$; $B = C_2H_5CH{=}CH_2$; $C = (CH_3)_2C{=}CH_2$.
Chapter 16 (p. 219): 9. $A = C_6H_5COOCH_3$; $B = CH_3C_6H_4COOH$; $C = C_6H_5CH_2COOH$; $D = C_6H_4(COOH)_2$. 10. 74; 0.612 g; C_2H_5COOH; $C_2H_5COOC_2H_5$. 11. $A = (COOCH_3)_2$; $B = CH_3COOC_2H_5$; $C = CH_3CHBr_2$. 12. $C_2H_5COOCH(CH_3)_2$.
Chapter 19 (p. 241): 8. (a) C_4H_{10}. 9. C_6H_7N. 13. $B = CH_3CH_2CH_2CONH_2$; $C = CH_3CH_2CH_2NH_2$; $D = CH_3CH_2CH_2OH$; $E = CH_3CH_2CHO$; $F = (CH_3)_2CHOH$; $G = (CH_3)_2CO$.
Chapter 20 (p. 247): 4. $A = CH_3C_6H_4CH{=}CH(CH_3)_2$; $B = CH_3C_6H_4CHO$; $C = (CH_3)_2CO$; $D = CH_3C_6H_4COOH$; $E = CH_3C_6H_4CONH_2$; $F = CH_3C_6H_4NH_2$. 5. 41 g; $A = CH_3CONH_2$; $B = CH_3CN$. 9. $A = CH_3CONH_2$; $B = CH_2CNCOOH$; $C = C_6H_5NHCOCH_3$.
Chapter 21 (p. 255): 6. 75 g; $A = CH_2NH_2COOH$; $B = HOCH_2COOH$; $C = (COOH)_2$.

INDEX

(The most important reference is given in bold type.)